THE WISDOM OF SCIENCE

# THE WISDOM OF SCIENCE

## its relevance to Culture and Religion

HANBURY BROWN, AC, FRS
*Professor of Physics (Astronomy) in the University of Sydney*

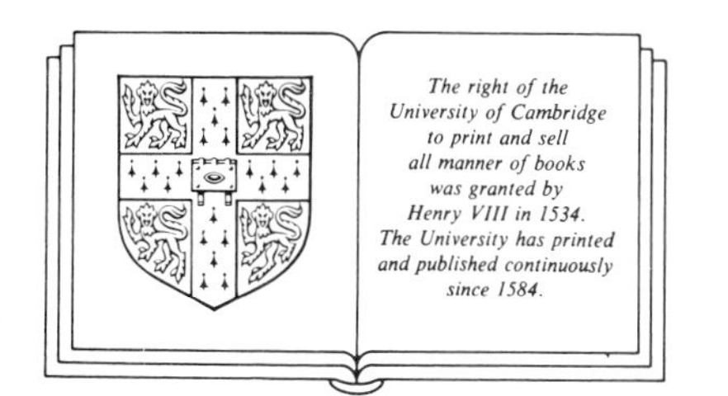


CAMBRIDGE UNIVERSITY PRESS
*Cambridge*
*London New York New Rochelle*
*Melbourne Sydney*

Published by the Press Syndicate of the University of Cambridge
The Pitt Building, Trumpington Street, Cambridge CB2 1RP
32 East 57th Street, New York, NY 10022, USA
10 Stamford Road, Oakleigh, Melbourne 3166, Australia



First published 1986

Printed in Great Britain by Billing & Sons Limited, Worcester

**British Library Cataloguing in publication data**

Brown, Robert Hanbury
The wisdom of science.
1. Science – History
I. Title
509 Q125

**Library of Congress Cataloguing-in-publication data**

Brown, R. Hanbury (Robert Hanbury)
The wisdom of science.
Bibliography
Includes index.
1. Science–Philosophy. 2. Science–Social aspects.
3. Science–Moral and ethical aspects. I. Title.
Q175.B7946 1986 501 85-17469

ISBN 0 521 30726 0 hard covers
ISBN 0 521 31448 8 paperback

# Contents

# Preface

For many years I ran an astronomical observatory in the Australian bush – Narrabri Observatory in New South Wales – and in that time I must have shown thousands of people round our installation and tried to answer their questions about what we were doing. The tour usually lasted about half an hour and the visitors were of all ages, nationalities and educational backgrounds. The work of the Observatory was highly technical; we were engaged in measuring the angular diameters of the bright visible stars using the newly invented technique of intensity interferometry. Even at the best of times, given facilities for projecting slides, a blackboard and an audience of physicists and astronomers, it is not easy to explain in half an hour how an intensity interferometer works. For that reason all I could do in the time was to explain to our visitors that our observations were important to a better understanding of the structure of stars, and that to understand stars is of considerable importance to astronomy.

What I really wanted to do was to complete the story by telling them that astronomy is an integral part of science, and that science is an integral part of modern civilization; but there was never enough time. To me it was profoundly unsatisfactory to send people away from this beautiful, highly sophisticated, and yet apparently useless instrument, surrounded as it was by thousands of sheep and a vast expanse of wheat, without having shown them that it was really part of an even greater world of which most of them were unaware, the ancient and invisible college of science.

In the years since the work of the Observatory was completed I have tried to write down what I would have liked to have said to my many visitors, if only there had been enough time. I have done this in three books; in the first (*The Intensity Interferometer*, Chapman and Hall, 1974) I have explained in great detail how the intensity interferometer came to be invented and how it actually works; in the second

(*Man and the Stars*, Oxford University Press, 1978) I have explained the relevance of astronomy to many things including science; and in the present book I have tried to show the relevance of science to our culture and religion.

Some years ago I heard a radio talk on religious affairs in which the speaker, a prominent Churchman, said that in his experience of answering questions about religion he had encountered three distinct types of people; firstly, there are those who are all for religion; secondly, there are those who are all against religion; and thirdly there are those who take religion seriously! Perhaps I should warn my readers that the discussion in this book of the religious dimension of science is intended simply and solely for people in this third group.

We live in a culture which, while dependent on science for its material welfare, is largely ignorant of the new ideas and perspectives, some quite startling, on which that science is based. Like the visitors to Narrabri Observatory, our society knows so little about science that, apart from the enquiry as to how much it costs, the only question usually asked of scientific research is – what use is it? That is a reasonable, if shallow, question and in my view scientists should be able to answer it, not only for their own particular branch, but for science in general.

In this book I have done my best to answer that same enquiry, but in a much deeper sense than the question itself implies. I have outlined what I believe to be the proper 'uses' of science and I have called my answer the 'Wisdom of Science' because I believe that the most valuable 'use' of science is in the getting of wisdom.

Hanbury Brown

# Acknowledgements

We wish to acknowledge the contributions made by the following organisations and individuals who kindly provided us with photographs: by permission of the Syndics of Cambridge University Library from Acton. c.51.24, p. 5; The British Library, p. 48; reproduced by Courtesy of the Trustees of the British Museum, 51; photograph from the Hale Observatories, 126; David Higham Associates, 10; Mary Evans Picture Library, pp. 2, 8, 18, 30, 65, 102, 104, 118, 132, 156, 173; John Milligan, from *Saturday Review, 1971*, p. 130; Museum of the City of New York, p. 28; NASA, pp. 36, 67, 175; National Optical Astronomy Observatories, p. 91; National Maritime Museum, London, p. 26; *New Scientist*: Cartoons by David Austin, p. 159 (16.10.80), p. 168 (9.8.84); PHOTO CERN, p. 72; Royal Greenwich Observatory, p. 67; Royal Observatory, Edinburgh, p. 67; reproduced by permission of the Trustees of the Science Museum, London, pp. 10, 16, 23, 39, 53; courtesy of Professor T. T. Tsong of Pennsylvania State University, p.57; University of Cambridge, Cavendish Laboratory, Madingley Road, Cambridge, England, p. 69 (i); C. E. Wynn-Williams at the Cavendish Laboratory, Cambridge, p. 70 (ii); U.S. Air Force Photo, p. 107; cover illustration: Bond of Union, lithograph, 1956, by M. C. Escher, © M. C. Escher Heirs, c/o Cordon Art – Baarn – Holland.

# 1 Changing the World

'The philosophers have only interpreted the world in various ways, the real task is to *change* it.'
Karl Marx

## 1.1 **Early attitudes to Science**

If we could go back to early medieval times in Europe we should be surprised to find how overpoweringly bookish the learned world was and how it lacked an idea which we now take for granted, the idea of progress. Scholars spent much of their time interpreting and comparing texts in old books and, in the case of the natural sciences, trying to reconcile what they read with the works of ancient authorities such as Aristotle or Galen. There was only a limited contact between ideas and reality and a discussion of topics in natural science was more likely to start from a belief based on authority than from an observed fact. Outside alchemy there was little practical work; an understanding of the natural world was not seen as something which grows by experiment, but as something which had been inherited from ancient Greece.

Thus, to most medieval scholars, the pursuit of science was not so much a bold adventure into the unknown as a search in the library for something which was already known and written down in the past. If the greatest minds in history – Aristotle and St. Thomas Aquinas – had explained all that it was proper or necessary to know, why look any further? It was not the actual quest for truth which attracted scholars, they wanted final answers; it was the product, not the process, which would give them joy. As St. Thomas Aquinas said, 'the final happiness of man consists in the contemplation of truth'.

Most scholars of those days were attached to the Church as clergy, monks or later, as University staff. Indeed, before the 16th century there were very few books, and not many of them were on science. In the revival of classical learning the translation of works of science had a low priority compared with works of theological or literary interest; Ptolemy's *Almagest* for example, one of the major scientific works of the past, was not printed in Latin until 1538. Besides, there were very few people outside the Church who could read; in Great

Britain the major expansion of education for the laity took place in the 16th and 17th centuries, during which time the number of schools open to them increased at least tenfold. Also in the early 16th century, very few of the books which did exist were in private hands; it has been estimated that, at that time, there were less than one hundred members of the upper classes in England who could claim to own one hundred books.

Thus, learning was a virtual monopoly of the Church and an active interest in science was rarely part of the saintly ideal. It was of course religion, not science, which really interested the learned world. After all, their principal task was to clarify man's relation to God, not his relation to Nature, and for that reason the natural sciences were valued more as allegories of theological truths than for their practical application. One of the greatest authorities of the Christian Church, St. Augustine, made it perfectly clear that one 'who can measure the heavens, number the stars and balance the elements' is no more pleasing to God than one who cannot, and that scientific knowledge was more likely to encourage pride than to lead people to God. Salvation was the goal, not material progress; science was not only superfluous to that aim but might even be dangerous. A thousand years after St. Augustine the message was still much the same; thus, in the 14th century

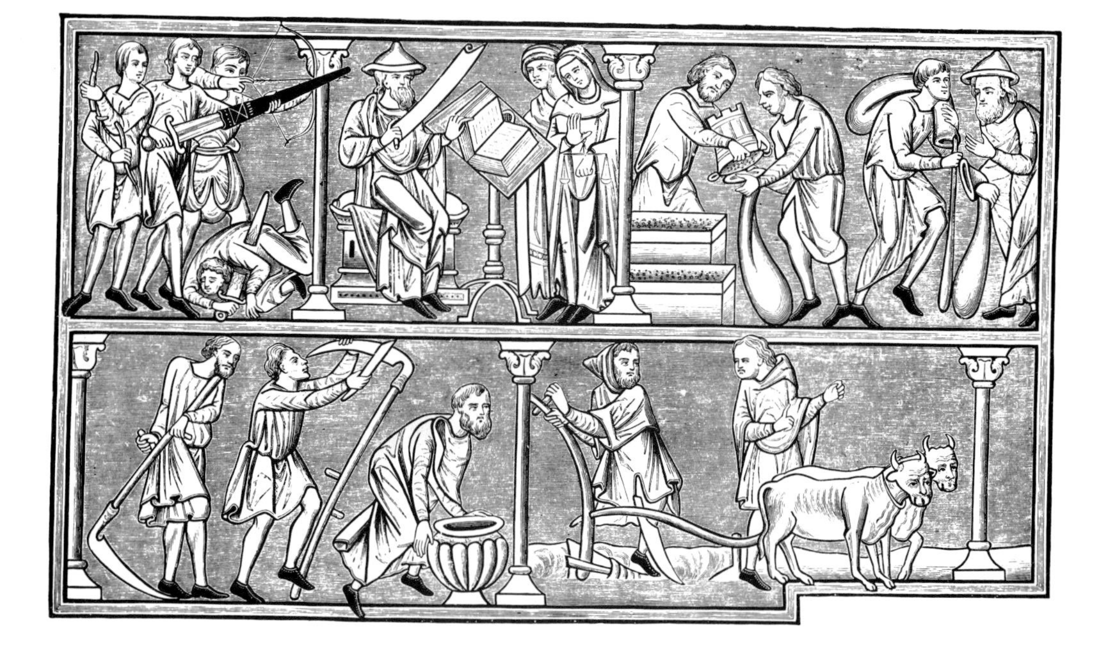

13th-century psalter showing War, Science, Commerce and Agriculture. Note that Science is presented as a passive activity which involves reading books.

Dante wrote in his *De Monarchia*: 'The human race is ordered for the best when according to the utmost of its power it became like unto God.'

Such beliefs did not suppress science but they certainly limited what it was permissible to find. Thus, in that classic and beautiful book on human anatomy – *De Humani Corporis Fabrica,* 1543, the first text-book to be based on human dissection – Andreas Vesalius carefully avoided a discussion of the doctrine that the heart is the seat of the human soul. He wrote: 'Lest I come into collision here with some scandal-monger or censor of heresy, I shall wholly abstain from consideration of the divisions of the soul and their locations.'

This monopoly of learning by the Church was broken in the 16th century largely by the rapid development of printing. By AD 1500 there were over a thousand printers in Europe and several million books. The economic develop-

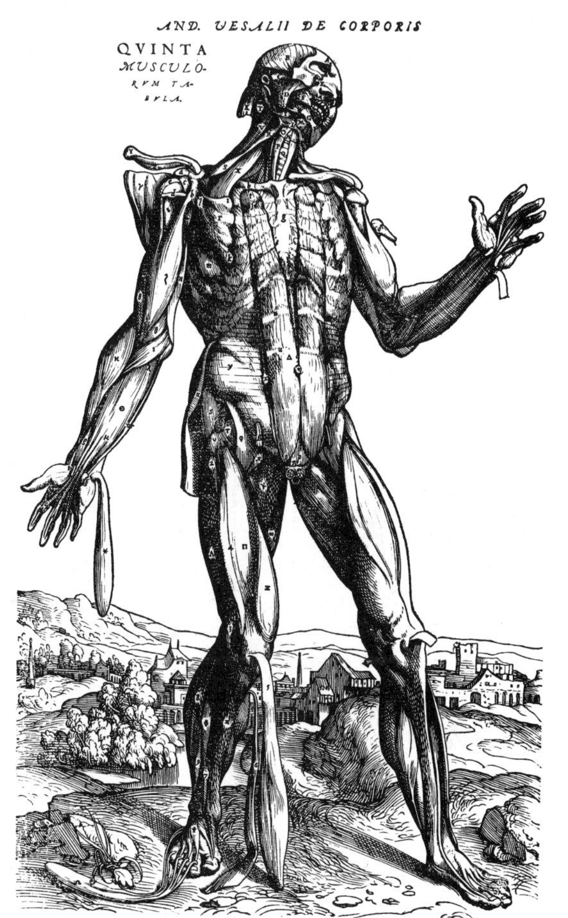


The muscles of the human body illustrated by Andreas Vesalius in the first text-book of anatomy to be based on the results of human dissection. *De Humani Corporis Fabrica*, 1543.

ment in Europe was starting to produce a society in which a prosperous merchant class was inclined to see progress more in terms of material than spiritual gain. Furthermore, the humanist scholars of the Renaissance were more interested in Man and this world than in God and the next world. In due course, these changing attitudes were to lead to an interest in investigating the world around us and so to the development of modern science, but in the 16th century experimental science was not yet seen as an important factor in the progress of society; the learned world still thought in terms of progress through 'literary' wisdom. This is not surprising because at that time there was very little scientific knowledge, and what little existed was of very little practical use. The major technical advances of those days were in mining, the treatment of metallic ores and in the working of glass and metal – activities which did not illustrate the value of science because they owed more to the empirical experience of craftsmen than to any systematic attempt to understand nature.

Before science could develop to the stage where it was capable of making an effective contribution to the welfare of society there had to be a strong social motive for pursuing it, as well as better scientific methods and, of course, better tools. Would-be scientists needed to be freed from the constraints of the scholastic tradition and to make contact, not only with each other, but with the practical world of arts and crafts. Above all, they needed to discover that the real art of making progress in science is to ask the right questions about nature. As we now know, the right questions are not metaphysical, but questions which can be answered by observation, experiment and mathematical analysis.

### 1.2 A new Vision – Science as the Key to Progress

A major prophet of the practical value of science and technology to the welfare of society appeared early in the 17th century. Francis Bacon's grand design was to 'restore and exalt the power and dominion of man himself, of the human race, over the Universe'. For him, the purpose of all knowledge, including the natural sciences, was power – the power to improve life on Earth by useful inventions. He wrote:

> 'Now among all the benefits which could be conferred upon mankind, I found none so great as the discovery of new arts, endowments, and commodities for the bettering of man's life.'

In Bacon's view the practice of magic is admirable in so far as it seeks to better man's life by dominating and improving nature; nevertheless he disapproved of it, not simply because he considered it to be fraudulent, but also because it is neither progressive nor cooperative. Bacon believed that people should work together to achieve a better future through applied knowledge; he saw science as being essentially a social activity.

Bacon derided the science of his day as being of very little practical use, and remarked of contemporary scholars that 'the last thing anyone would be likely to entertain is an unfamiliar thought'. In his view the chief failing of scholars was that they substituted talk for experiment and contemplation for action. They were easily satisfied with purely verbal solutions to real physical problems; they confused science

Francis Bacon (1561–1626).

with religion and were always seeking final causes and doctrines which would explain physical phenomena once and for all. Reacting against this tradition, Bacon proposed that the truth and value of knowledge should be tested by its utility; hence the worth of any system of philosophy should be judged by its contribution to the welfare of humanity. With one eye on the Establishment, he argued that to advance human welfare was to serve God; both faith and science should be judged by their works.

If science was really going to advance without the help of ancient authorities, then there was a need for a new method of making progress. Bacon put forward a 'new method' (*Novum Organum*) based on the idea that scientific knowledge is cumulative, and that it can be increased with time by methodical hard work. He visualised teams of people carrying out a multitude of experiments planned to embrace all possible enquiries which are relevant to the welfare of society. These experiments would yield a great mass of facts from which new, more general, laws of nature would be extracted by a process which philosophers call induction. To illustrate his plan he wrote, like H.G.Wells, a science fiction story (*The New Atlantis*) about an ideal society on a remote island. This society was remarkably conventional and, unlike H.G.Wells, he made it Christian. At the time of the crucifixion it had experienced a separate revelation of Christanity; through the agency of St. Bartholomew a small ark, marked by a pillar of fire, appeared about a mile off-shore bearing a complete copy of the Bible which included several books of the New Testament which, by the way, would not have been written at that time.

The novel thing about Bacon's utopian society was that it applied science methodically to the welfare of society. There was on the island a large establishment called Salomon's House which was devoted entirely to experiments in applied science, encompassing every practical application of which Bacon could conceive. He described in detail how this research was staffed and organized and he used it to illustrate how his 'new scientific method' could be put into practice. In retrospect we can see that Bacon's 'new method' was not really a practical way of doing science; rather unkindly it has been called 'research by administration', which, after all, is only to be expected from a man who was at one time Lord Chancellor.

Francis Bacon is not someone picked out of the history of science because his views happen to agree with what we think nowadays. In fact, what he wrote and said was currently influential and has survived, not simply because he said it well, but because he said it at the right time. In his day there was a rapidly growing interest in science, and by linking the purpose of the natural sciences firmly to their practical applications, Bacon gave society a powerful motive for pursuing science. Furthermore he supported it by the novel and refreshing argument that we are more fortunate than our ancestors because we are heirs to all that has been discovered. For Bacon, what we think of as antiquity was the youth of the world; this was a novel point of view because in his day it was quite common to argue that the world had passed its peak and was already senile.

To be sure, much of what Bacon said had been said before, but no-one had directed attention so forcefully to the supremacy of fact and to the weakness of medieval thought; furthermore, no-one had painted such a clear picture of the possibilities of science as a forward-looking, beneficent, co-operative, social activity. Bacon foreshadowed the large modern laboratory working on applied science. He is, so to speak, the patron saint of the utilitarian view of science. Indeed he might even be called a martyr; while driving near Highgate in the winter of 1626 he got out of his carriage to stuff a dead chicken with snow in an effort to find out if the cold would preserve it, and caught his death of cold.

## 1.3 **New Tools and Methods for Science**

At the time Bacon wrote – the early 17th century – there was good reason for optimism about the future of science and technology. Scientists were making rapid progress largely due to the new scientific instruments – the telescope, microscope, thermometer, barometer, pendulum clock and the air pump – which appeared for the first time in practical form.

Histories of science are often written in terms of outstanding people like Newton and Einstein so that they give the impression that the progress of science depends largely on the development of new theories. It would be nearer the truth to say that it depends largely on the development of new instruments, and hence on new materials and new ways of making things. The progress of astronomy, for example, owes more to the invention of the telescope than to any other single factor; likewise biology and medical science would not have got very far without the microscope. Both of these important tools were invented by craftsmen, not by scientists, and both appeared in a practical form at the beginning of the 17th century; their development, in turn, depended on the progress of metallurgy and the manufacture of glass in the 16th century.

The telescope appeared in Holland in 1608 and its invention is usually credited to a spectacle maker, Hans Lippershey of Middleburg. As one story goes he invented the telescope by accident while looking at a church steeple with two spectacle lenses, one in each hand. Whatever may be the truth of that story, there is no doubt that the same discovery had been made several years before, but by people who failed to recognize its practical importance.

Astronomers were quick to realize the importance of the telescope and within a year of its appearance in Holland there are records of its use in astronomy in at least three countries, England, Germany and Italy. In Italy Galileo, on hearing of the invention, made his own telescope and drew the attention

Venetian Senators see the satellites of Jupiter through Galileo's telescope.

of the Venetian Republic to its potential importance in marine affairs. It proved to be of immediate practical use at sea, particularly for the navigation of ships near the coast. Galileo went on to apply the telescope to astronomy and in so doing gave us a new view of the sky. Through his telescope he saw satellites circling Jupiter, mountains on the Moon, spots on the Sun, the phases of Venus, myriads of stars in the Milky Way, and many other new things. As we shall discuss further in §2.2, his observations weakened the hold of the orthodox cosmology of Aristotle and Ptolemy and helped to establish an entirely new model of the solar system, the heliocentric model put forward by Copernicus.

The microscope also appeared in a practical form at about the same time and place. At first, the best instruments consisted of a single tiny lens of good shape and very short focal length and were developed by an Amsterdam merchant called Anthony van Leeuwenhoek. With his microscope van Leeuwenhoek and other microscopists, among them Robert Hooke in England, discovered a whole new world, the world of the very small. For the first time they saw things which are

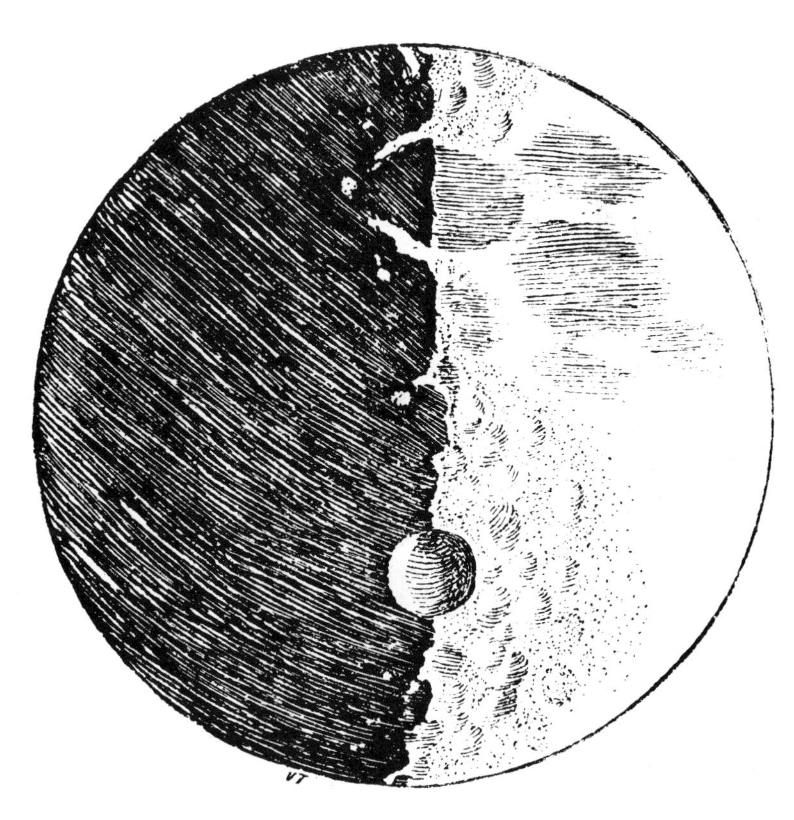

Galileo's sketch of the Moon as he saw it through his telescope. *The Sidereal Messenger*, 1610.

fundamental to our understanding of nature such as protozoa, bacteria, spermatozoa, red corpuscles and the capillaries which carry blood from the arteries to the veins and so on. To illustrate how important these discoveries were, and how intimately the advance of science depends on new instruments, we have only to recall that with his microscope, van Leeuwenhoek established the life cycle of the weevil and the flea, and thereby undermined the medieval theory that they were 'spontaneously generated' by wheat and dirt. As another example Robert Hooke established with his microscope that cork, and some other vegetables, have a cellular structure and he gave us the name 'cells'. The development of the mechanical air-pump was the first step towards understanding the nature of a gas and the fundamental part which air plays in respiration and combustion. Its

Anthony van Leeuwenhoek using his microscope (1680).

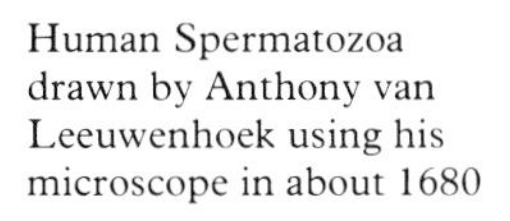

Human Spermatozoa drawn by Anthony van Leeuwenhoek using his microscope in about 1680

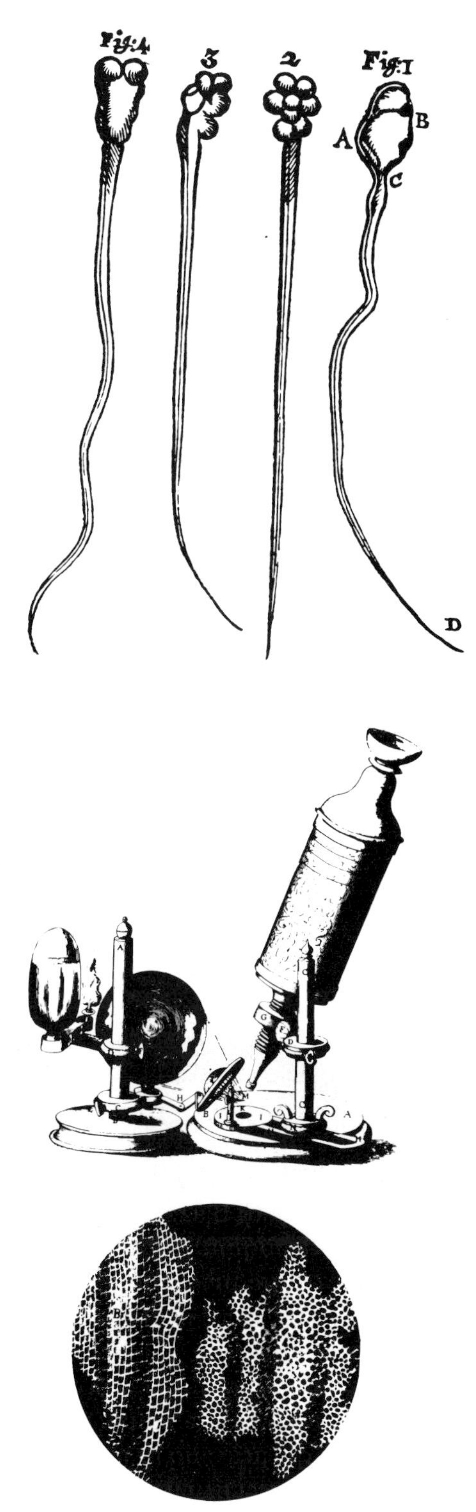


Robert Hooke's microscope (1665) together with an impression of the cellular structure of cork which he discovered with the instrument.

invention is usually credited to Otto von Guericke who exhibited it in 1654.

The thermometer, barometer and pendulum clock made it possible to make precise and objective measurements of the basic physical parameters of heat, pressure and time. The fact that Galileo used the beat of his own pulse as the unit of time in some of his classic experiments on motion, reminds us of the limitations imposed on early science by the lack of measuring instruments! The techniques of mathematics were also advanced in the 17th century, making it increasingly possible to turn qualitative speculation into quantitative science and to tackle problems which could not be solved in words. For example, in Great Britain, Newton could explain the motions of the solar system in terms of the mutual attraction between large and massive bodies because he had invented, but not published, his 'method of fluxions' or, in other words, the calculus. In addition to the calculus the 16th and 17th centuries saw the introduction of modern algebraic notation, decimals, logarithms, and co-ordinate geometry. It is hard nowadays to imagine doing scientific research without them, and it makes one realize how much we take for granted!

So much for the physical tools of science, but what about advances in scientific method? As we have seen, it was Bacon's ambition to advance science by improving the actual method of doing it, and much the same thing can be said of the great 17th century philosopher and mathematician René Descartes. But while Bacon advocated a massive programme of experiments followed by the induction of general laws, and failed to appreciate the importance of mathematics, Descartes did roughly the opposite. For him mathematics should explain all that can be known about order and measure; he regarded it as the essential pattern of science and with it he aimed to build a science of nature which would have the same certainty as the theorems of mathematics. His principal tool was doubt. Starting with simple assumptions (e.g. 'I think, therefore I am') which he considered to be indubitable, Descartes sought to deduce all the known phenomena of the natural sciences and the existence of God. He divided the world into matter and mind and considered the realities of the material world to be space, extension and motion. All the relations between these realities could be explained in terms of the motion of corpuscles interacting by physical contact, not by fields of force, and could be expressed by mathematical laws. Thus it was really Descartes, not Newton, who laid the foundations of what came to be known (§2.4) as the 'Mechanical Philosophy'.

However, in real life, people don't learn how to do scientific research or how to make discoveries by reading about scientific method in books. The most important things about scientific method are usually learned by example, by following someone who knows how to do research. For that reason, it seems likely that a brilliant, productive and practical scientist like Galileo had more influence on the way in which scientific work was actually done than either Bacon or Descartes.

Galileo is best known for his application of the telescope to astronomy by which he demonstrated, so dramatically, the importance of observational fact; but he did much more than that. By his work on the science of mechanics, particularly on accelerated motion, he showed how, in physical research, one must start by looking for quantities that can be measured and then try to establish the relation between them. He showed how to combine observation, experiment, speculation and mathematical analysis in a way which is typical of modern science. In retrospect he is often called 'the first modern scientist'.

Nevertheless, it was Isaac Newton who gave the world the most powerful, convincing and influential demonstration of this new scientific method. In his remarkable book, *The Principles of Mathematical Philosophy,* 1687, he completed the cosmological revolution started by Copernicus, Kepler and Galileo by showing mathematically how all the motions which we observe in the solar system, from the planetary orbits to the ocean tides, can be explained and predicted in terms of a few simple physical laws. By so doing, he demonstrated how observational data, speculative theory and mathematical analysis can be combined to solve an extraordinarily difficult problem. In other words, he showed the world how to do modern science.

Newton's work is essentially modern in the way he used observational data to verify his theoretical analysis and in the way he accepted and made use of the mysterious concept of gravity without postulating its 'final cause'. It is also essentially modern in its use of mathematics. Newton proved, for example, that the gravitational attraction of a large sphere, such as the Earth, acts as though the whole mass of the sphere is concentrated at its centre, and he showed that Kepler's laws of planetary motion are a consequence of the inverse square law of gravitational attraction. Both these proofs are fundamental to our understanding of celestial mechanics and neither of them could have been accomplished entirely in words; they depended on the use of mathematics. By the end of the 17th century physical science had emerged as a separate intellectual discipline – distinct from literature, theology and history – with its own special language, mathematics.

### 1.4 Science forms a Community

If science was to develop its own identity, free from the constraints of the Church and scholastic world, then there was a need for scientific societies to give a focus to the scientific community. The first attempt to establish a scientific society was unsuccessful. The Academia Secretorum Naturae was founded in Naples in 1560. Not surprisingly the name proved too much for the public, who suspected it of black magic. It was duly suppressed and the President was ordered by the Pope to abstain in the future from illicit arts. The next attempt was more successful, the Academia dei Lincei (Lynxes) was founded in Italy in 1603 and some years later enlisted Galileo as a member.

In England the Restoration of King Charles II brought science into fashion and the Royal Society was founded in 1660; two years later it was granted a charter, and very little else, by the King. Its foundation and early devotion to experimental work echoed Salomon's House in Bacon's *New Atlantis*. In France science was more official; the Académie Royale des Sciences was created by Louis XIV in 1666. Unlike the Royal Society, where Fellows have always had to pay a subscription, the French Academicians were granted a pension and their Academy was given adequate funds which, inevitably, entailed quite onerous obligations to the State.

Nowadays we take our scientific view of the world so much for granted that we are apt to forget how much it owes to a few great explanatory ideas, such as the atomic theory of matter and the theory of evolution, which unify and make sense of what our many specialized branches of science tell us. However when the early scientific societies explored the natural world they took all science as their agenda, and they did so without the benefit of those great simplifying principles which allow us to relate one phenomenon to another and so to make sense of the world around us. The sheer diversity of natural phenomena must have been bewildering, let alone the difficulty of explaining them. I often wonder how those early enthusiasts preserved their faith in science.

Consider, for example, a list of the topics which engaged the attention of the Royal Society of London in its first few years. Thomas Spratt, writing in 1667, tells us that in its first few years the Society busied itself with experiments under 11 major headings:

'Fire and Flame, Air, Water, Mines, Metals, Ores and Stones, Gravity, Pressure, Levity, Fluidity, Firmness, Congruity, etc., Light, Sound, Colours, Taste and Smell, Motion, Chymical Mechanical and Optical.'

Furthermore, he tells us a good deal about the programme of experiments under each heading. Viewed as a programme for a modern scientific society it was incredibly diverse and some of it was positively alarming. Imagine, for example, experiments with a 'Poysoned Indian Dagger on several Animals', or on 'Feeding a Carp in the Air'. As far as science is concerned these meetings may have been bewildering, but at least some of them may have been entertaining!

These early scientific societies gave science an identity and put people who were interested in science in touch with those who were interested in applying it. They did this partly by holding meetings at which people with different interests could meet and exchange information, and partly by publishing scientific journals. The *Journal des Sçavans* appeared in France in January 1665; two months later the *Philosophical Transactions* was published by the secretary of the Royal Society, Henry Oldenburg, initially as a private venture. The *Philosophical Transactions* was the first journal devoted exclusively to science and, by-the-way, was the first scientific journal to publish 'scientific papers' signed by their authors. The knowledge on which modern science is built is essentially cumulative and public, and it was largely the scientific journals which made this possible. Through these journals people could keep in touch with science, both at home and abroad, and they could make their own contribution to its growth by publishing an account of some experiment, even something modest, which they had done themselves.

Fig. M.

## 1.5 **Early Science as seen by the Public**

It is interesting to look at some of the early public reactions, some hostile, some friendly, to the growing interest in science and to the activities and pretensions of the new scientific societies. Thomas Spratt in his *History of the Royal Society,* 1667, which was written largely as a justification and defence of the Society, identified some of the sources of hostile criticism. He wrote:

'...while some over-zealous Divines do reprobate Natural Philosophy, as a carnal knowledge and as too much minding worldly things, the men of the World, and business, on the other side, esteem it meerly as an idle matter of Fancy, and as that which disables us from taking right measures in humane affairs. Thus by one party it is censured for stooping too low, by the other, for soaring too high....'

Clearly there have always been difficulties in maintaining a satisfactory balance between 'pure' and 'applied' research!

There were also reactions from men of letters and high society, some of whom actually attended the meetings of the Royal Society and read the *Philosophical Transactions.* We find, for example, the famous philosopher John Locke, a Fellow of the Society and the leading exponent of empiricism, telling us in his great *Essay Concerning Human Understanding,* 1690, that 'natural philosophy is not capable of being made a science'. He goes on to say:

'Experiments and historical observations we may have from which we may draw advantages of ease and health and thereby increase our stock of conveniences for this life; but beyond this our talents reach not...'

Judging from the popularity of a comedy which was staged in London in 1676, many people must have felt that the scientists of the Royal Society were failing, ludicrously, to produce these promised 'advantages of ease and health.' The play was called *The Virtuoso* and was written by Thomas Shadwell. In those days a Virtuoso was someone who was interested in science or in what were then considered to be scientific pursuits, such as collecting curiosities. The principal character in the play, Sir Nicholas Gimcrack, is a scientist and a collector, and Shadwell uses him as a tool to poke fun at scientists in general and at the Royal Society in particular. Thus when Sir Nicholas claims to read his Geneva Bible by the light of a putrid leg of pork, Shadwell is making fun of the apparent uselessness and diversity of the experiments reported to the Royal Society and, in particular, of

*Left*
Enthusiasm for Science in the 17th-century home; man and woman operating a brass sextant (1658).

some recent observations by Robert Boyle of natural luminescence.

Again Shadwell pokes fun at scientists for not pursuing the practical applications of their researches. Thus Sir Nicholas learns to swim by imitating a frog, but he does it on a table because, as he says, he hates water:

'I content myself with the speculative part of swimming; I care not for the practick, I seldom bring anything to use; tis not my way. Knowledge is my ultimate end.'

Finally Shadwell caricatures the perennial concern that the advance of science will put people out of work. Sir Nicholas invents a loudspeaking device – a 'stentrophonical tube' – with which one person can preach to the whole country, and points out that this would enable the King to take all the Church lands into his own hands because one parson, the King's chaplain, could do the whole job. When asked what will happen to the out-of-work parsons, he suggests that they should make fishing nets.

A professor at Lagado working on a project to substitute spiders' webs for silk.

Turning from the stage to books, we find Jonathan Swift also mocking the uselessness of Science in two popular books, *The Tale of a Tub* and *Gulliver's Travels*. The 'Academy of Projectors' at Lagado in *Gulliver's Travels* is a satire on Salomon's house in Bacon's *New Atlantis* and on the early Royal Society. The Academy at Lagado was established by royal patent, just like the Royal Society, and was devoted to the improvement of all 'arts, sciences, languages and mechanics'. The professors of the Academy 'contrived new rules and methods of agriculture and building and new instruments and tools for all trades and manufactures'; in other words, its aims were those of Bacon, to apply science to the welfare of society. Gulliver makes a tour of some of these 'practical' projects which are all, without exception, ludicrously impractical. One professor is trying to extract sunbeams from cucumbers; another to develop new methods of building houses starting with the roof and working downwards; another to separate human excrement into its original food, and so on. Swift portrays scientists not only as being ludicrously impractical, but also as money-grabbers, makers of false promises and as personally repulsive. Attacks on science are still being made to-day and scientists still promise more than they can do. But the scientists of modern fiction are no longer figures of fun, nor are they accused of doing too little; nowadays they are more likely to be seen as sinister and to be accused of doing too much.

## 1.6 **Early attempts to make use of Science**

Why did the members of the Royal Society in the 17th century pursue science? Was knowledge their 'ultimate end', as Sir Nicholas Gimcrack says, or did they, with Francis Bacon, hope for material benefits? Bacon in his eloquent propaganda for the application of science had recognized that there must first be 'experiments of light' to discover the causes of things, and then there must be 'experiments of fruit' to apply this knowledge to practical ends. An analysis of the programme of the early Royal Society suggests that the majority of its members, perhaps two thirds, were more interested in gaining 'fruit' than 'light'. They sought to apply science to practical affairs, to mining, marine navigation, land surveying, military arts such as gunnery, the textile industry, and so on.

Even so, although painstaking studies were made of

various crafts and industrial processes and although the economic climate favoured the application of science to industry, the practical results were rather disappointing. It took far more time and energy for applied science to produce tangible benefits than anyone had expected. Indeed, the major scientific achievements of the 17th century were, in modern jargon, more 'curiosity oriented' than 'mission oriented'; science itself made great steps forward but its applications did not. In retrospect this is not surprising; quite simply the scientists of those days didn't know enough to tackle anything except the simplest problems. Many of the more important problems which they tried to solve proved to be too difficult, and that is why so many of the things which

The hazards of medicine and surgery in the 17th century. Science brought little improvement to the actual treatment and cure of diseases until the 19th century.

they did seemed to the public to be trivial if not ridiculous. As any consulting engineer will confirm, the attractive notion that old crafts and manufacturing processes can be improved by a little science is, more often than not, very difficult to put into practice; it usually turns out that one needs to know a great deal more about science than would appear at first sight.

This was particularly true of medical science. Despite the great scientific advances of the 17th century, such as the discovery of bacteria and the circulation of the blood, there was little improvement in the actual cure and treatment of diseases until the 19th century. With hindsight, it seems likely that the main value of all this early enthusiasm for science was not to impart scientific knowledge but to encourage scientific attitudes in industry, and to draw attention to the need to study systematically the basic facts of physics and chemistry rather than the exotic curiosities of nature. By the end of the 17th century science had become a serious matter, less of a hobby for the upper classes and more of a profession for the middle classes; it was a time of transition from the amateur to the professional.

## 1.7 **Science meets Industry and is seen to be useful**

Although Newton's *Principia* demonstrated that science can lead to a better understanding of the world, it was left to the scientists of the late 18th and 19th centuries to confirm Bacon's forecast that it could be used to advance our material welfare. One of the principal factors which brought this about was the foundation in the 18th century of provincial scientific societies which put scientists in much closer touch with industry than did the Royal Society in London. In Great Britain, 'philosophical societies' – in those days science was called natural philosophy – were founded in all the major centres where the Industrial Revolution was taking place.

One particularly influential society, whose origins can be traced back to 1750, was the Lunar Society of Birmingham, so-called because its meetings were held on the monday nearest to full Moon so as to make travel easier after dark. Its membership in 1766 reads like a *Who's Who* of 18th century progress; Josiah Wedgwood was a potter interested in chemistry; James Watt was an engineer with an academic interest in physics; Matthew Boulton was a manufacturer of metal parts who, together with James Watt, made steam engines; Joseph Priestley was a Unitarian clergyman with an active interest in chemistry and physics; Erasmus Darwin was a physician and a poet, and so on. Men like these appreciated the importance of scientific method and knowledge, and were interested both in applying science to industry and in making it available to the public. Indeed, the public lectures on science organized by these societies played a significant role in scientific education at a time when there was little or no science taught in the schools and universities.

It was in these provincial societies that scientists really came face to face with the problems of industry, which, in the 18th century, were largely chemical and arose in the major industry of that time, the textile industry. Although the investigation of chemical processes in the previous century had yielded a vast number of facts, there was as yet no systematic understanding of these data. While Newton had already clarified the understanding of the *mechanical* behaviour of matter, it took another century to reach a comparable understanding of its *chemical* behaviour.

As we shall see in §2.5 a revolution in the understanding of the chemical behaviour of matter started in the late 18th century, and the foundations of quantitative chemistry were laid by Lavoisier when he published his *Elementary Treatise on Chemistry* in 1789. By the end of the 18th century chemical science had advanced to the point where it could be applied with increasing success to the problems of industry. As a result, improvements were made to many processes of commercial importance such as the manufacture of sulphuric acid, soda and bleaching powder. And that was not all, other people besides the chemists were showing that science could be applied usefully to problems of practical importance.

The first useful steam engines were developed at the end of the 17th century to meet the urgent need for a better method of pumping water out of deep coal mines than the use of horses to hoist buckets. In Great Britain, the increasing demand for iron and the decreasing supply of wood as a fuel for the iron-makers' furnaces had led to the digging of deeper coal-mines, which, inevitably, were liable to be flooded.

Those early steam pumping engines owed little to scientific understanding until, in 1765, James Watt invented the condensing steam-engine. At that time Watt was employed as an instrument maker at Glasgow University and was a close friend of Professor Joseph Black who is well known for his discovery of latent heat. In the course of repairing an early

Newcomen engine, Watt was impressed by its prodigious waste of heat. He made a systematic study of the heat losses in the engine and, consulting Black, he related them to the physical properties of steam and to the conduction of heat by metals. He demonstrated that the major loss in the engine could be avoided by the use of a separate condenser which increased its overall efficiency by a factor of about four. This was the major breakthrough which ushered in the 'Age of Steam' and made the steam-engine the almost universal prime mover of industry and transport in the 19th century.

It may well be, as has been said, that 'science owed more to the steam-engine than the steam-engine owed to science'. To be sure, most of the early development was carried out by artisans who had no formal training in science. Furthermore there was very little known about heat in the 18th century; thermodynamics was essentially a 19th century creation. Nevertheless, the invention and development of the condensing steam-engine was an early and influential example of the application of science to one of the most important developments of modern industry – the replacement of water power,

A Boulton and Watts condensing steam engine of 1788. The first source of power which could be applied on a large scale to drive any machine, anywhere, at any time, needing only fuel and water. It set industry free from water, wind or animal power, and so made industrial expansion possible. Steam was condensed in a separate vessel from the cylinder.

horsepower and human labour by the steam engine. We are reminded of this by the now almost obsolete unit of power, the horsepower, which was introduced by Watt to describe the power of his new engines, and by the modern unit of power, the watt, which was named after him.

In the development of marine navigation, science played a more prominent part than in the development of the steam engine. When reading about the great voyages of exploration we are apt to forget that, before the 18th century, there was no way in which the position of a ship at sea could be fixed on the map once it was out of sight of land. The systematic measurement of latitude as an aid to navigation of ships dates from the Portuguese voyages of exploration in the 15th century, but even Prince Henry the Navigator failed to develop a method of measuring longitude. This was of course a limitation to the safety and effectiveness of the world's navies; more importantly it was a serious limit to the safety and profitability of merchantmen and to the growing international trade in raw materials for industry.

The importance of the problem, and the fact that its solution would involve advances in astronomy, was recognized in Great Britain by the foundation of the Royal Observatory in 1675 – an event, by the way, which marks the beginning of the 'scientific civil service'. To quote the warrant from Charles II, the Observatory was founded 'in order to the finding out of the longitude of places for perfecting navigation and astronomy'. Progress proved to be rather slow, which is not surprising in view of the fact that the King provided a building and a salary of £100 per year for the Astronomer Royal and nothing else. All the instruments and assistants had to be begged, borrowed or paid for by the Astronomer Royal himself – John Flamsteed.

Following a major disaster to the British fleet, due to poor navigation, the Government decided that, if the longitude of ships at sea was to be found, something more urgent needed to be done. In 1714 they established the enormous prize of £20,000 for any one who 'could discover the longitude at sea'. Within 50 years they had two practicable methods, the method of lunars and the marine chronometer.

The first practicable system, the method of lunars, was the product of remarkably advanced celestial mechanics and refined observational astronomy. In this method the angular distance between the Moon and a bright star was measured, and using this angle and the local time the longitude of the observer was then computed. To achieve a reasonable accuracy, say 1/2 a degree in longitude, the positions of both the Moon and the star had to be known, at any given time, with an accuracy of about 1 minute of arc. Before the 18th century such an accuracy was impossible. Although people had studied the Moon since time immemorial, it was not until Isaac Newton published his great work on motion and gravitation *Principia*, 1687 that it was possible to develop an accurate mathematical theory of how the Moon moves.

It soon transpired that the motion of the Moon is extremely complicated, and to develop the theory to the point where it could predict the position to within 1 minute of arc took more than 50 years and the efforts of many of the leading mathematicians of the day. The first accurate tables of the Moon's position were calculated in 1753 by a cartographer, Tobias Mayer, using methods put forward by the great mathematician, Leonhard Euler, in a prize essay written for the Academy in Paris in 1748. Following the publication of Mayer's tables, the method of lunars became generally available, and was used with great success by Captain James Cook in that model of perfect navigation, the voyage of the *Endeavour* (1768-1771).

As it happened, the working life of this new method of finding longitude was short. It was soon to be rendered obsolete by the work of a skilled craftsman, John Harrison, who developed a new type of mechanical clock which could keep accurate time in a ship at sea, the marine chronometer.

The development of navigation as a whole, was a particularly effective demonstration of the application of science to practical affairs because it involved so much apparently 'useless' science. For example, in a simple measurement of latitude, the position of a bright star or the Sun must be known with an accuracy considerably better than 1 minute of arc. Such an accuracy demands not only refined techniques of observation but an understanding of how to correct these observations for a number of quite obscure and complicated effects, such as the aberration of light, the precession of the equinoxes, the nutation of the Earth's pole and the refraction of light in the Earth's atmosphere. Furthermore, in observing stars from the deck of a ship it is necessary to correct the observations for the dip of the horizon due to the height of the observer above the sea, and this correction, although simple to calculate, involves knowing the size of the Earth. Again, in correcting the observations of the Moon for parallax, it is necessary to know the ratio of the distance of the Moon to the size of the Earth.

After a long history of attempts to find the position of a ship at sea, the problem was finally solved in the 18th century. Although the marine chronometer was the invention of a skilled craftsmen, not a scientist, nevertheless mathematics and physics were widely seen to have played a major part in

The first marine chronometer. John Harrison's first 'sea clock' (H1) which he completed in 1735. It was tested on a barge in the Humber and later on a voyage to Lisbon.

the solution. It was the first clear demonstration that 'modern' science could help to solve an important practical problem.

## 1.8 **Knowledge meets Power – the first Science-based Industries**

It was not until the mid-18th century that the Industrial Revolution in Britain showed what a profound effect advances in technology can have on everyday life. Broadly speaking the main result of that revolution was to transform much of Britain from a community of farm workers living in the country into a community of factory workers living in towns. Thus in 1750 there were only two cities with more than 50,000 inhabitants, but a century later there were 29; ever since then, more people have lived in the towns than in the countryside.

The principal technological advances which brought about these remarkable changes were the mechanization of industrial processes, such as spinning and weaving, the substitution of steam engines for human, animal and water power, and improvements in the mining and working of raw materials, most notably of iron. Thus the Industrial Revolution was a major demonstration of Bacon's thesis that, through the purposeful application of knowledge, it is possible to advance the material welfare of society. It was not, however, a demonstration that the discovery of new knowledge by scientific research can maintain and accelerate this progress. In fact, most of the technical advances on which the Industrial Revolution was based, although helped by science, were not critically dependent on recent scientific discoveries; to a large extent they made use of knowledge which had been lying about for years.

Not surprisingly it was the engineers, and not the scientists, who, in those days, were regarded as the principal authors of material progress. The great works of the 18th and 19th centuries, the canals, factories, railways, bridges and viaducts were public evidence of progress due to the skill and daring of the great engineers – men such as Telford and Brunel in Great Britain, Roebling and Ellet in the USA, and so on. If in early Victorian days we had asked the man or woman in the street to name a 'great engineer' they could probably have done so, but I doubt whether they would have been able to name a 'great scientist' – nowadays the situation would be reversed.

Indeed, the idea that the material progress of society is linked to the progress of science – one of the principal ideas which separate us from the medieval world – had not yet been illustrated with sufficient force to become part of the conventional wisdom. Although at the end of the 18th century science was seen by practical men as useful, and by philosophers as a paradigm of the use of reason, it was not yet seen as the principal agent of material progress. Admittedly it had proved useful in the development of the chemical industry and of marine navigation, but there was still not enough known at the beginning of the 19th century for science to become a major factor in the advance of industry. For

A major triumph of 19th-century engineering. The building of the Brooklyn Bridge, New York (1881).

example, although it had been established that some substances are elements and that some are compounds, there was no generally accepted theory of the atomic structure of matter; although quite a lot was known about static electricity, very little was known about electric currents and, in particular, their relation to magnetism had not yet been discovered; both heat and electricity were still believed to be fluids and, as yet, there was no useful concept of energy.

All these topics were greatly advanced during the first half of the 19th century. The atomic theory of matter was established following the work of John Dalton; electric currents were studied and their relation to magnetism was discovered; heat was shown to be a form of energy and a new branch of science, thermodynamics, was founded. By mid-century the foundations of what we now call 'classical science' had been laid; the stage was set, in Bacon's words, for 'knowledge to meet with power'.

The first convincing demonstration of the importance of scientific research to a major industry was the discovery, by William Perkin, of the aniline dye, magenta, in 1856. He made this discovery while trying to synthesize quinine in the laboratory of the Royal College of Chemistry in London. Perkin's discovery led to the foundation of the great synthetic dye industry and, in Germany, to the formation of some of the first industrial research laboratories. The idea of an organized scientific laboratory, in which a group of professional scientists is assembled to tackle some specific problem, originated in the universities; it was the German chemists who introduced it into the factory and created the first industrial research laboratories. The greatest invention of the 19th century, so Alfred North Whitehead tells us, was the invention of the method of invention.

Another even more conclusive demonstration of the usefulness of science was the foundation of an entirely new industry – the electrical industry. The phenomenon of static electricity had been known since the days of ancient Greece and many experiments with atmospheric electricity were made in the 18th century leading, most notably, to the installation of lightning conductors; but it was not until 1800 that the invention of the electric battery by Alessandro Volta made it practicable to experiment with electric currents. After that, it was not long before, in 1820, H.C.Oersted made one of the basic discoveries of physical science while giving a lecture! While he was demonstrating the electric current from one of Volta's batteries, he noticed that a compass needle, which happened to be on the lecturer's desk, deflected. He had found, entirely by accident, that an electric current produces a magnetic field.

Oersted's discovery lead directly to a new way of detecting electric currents and hence to a practical form of electric telegraph. Previous efforts to develop a telegraph based on the detection of a static electric charge had failed. The problem was now solved, and the first practical telegraph system appeared in 1837 and, in Great Britain, was immediately applied to the railways. By 1866 a submarine telegraph was operating between the USA and Great Britain and by 1872 all the principal cities of the world were linked by telegraph.

In the meantime, Michael Faraday had been hard at work in the laboratory of the Royal Institute in London. In 1831 he demonstrated the connections between electricity, magnetism and motion, which led directly to the development of the dynamo, the alternator and the transformer. Almost 50 years later, following the invention of electric lighting, Edison's Pearl Street power station was opened in New York; it had six

A completely new industry based on laboratory science. An electric power station in Brooklyn, New York (1890).

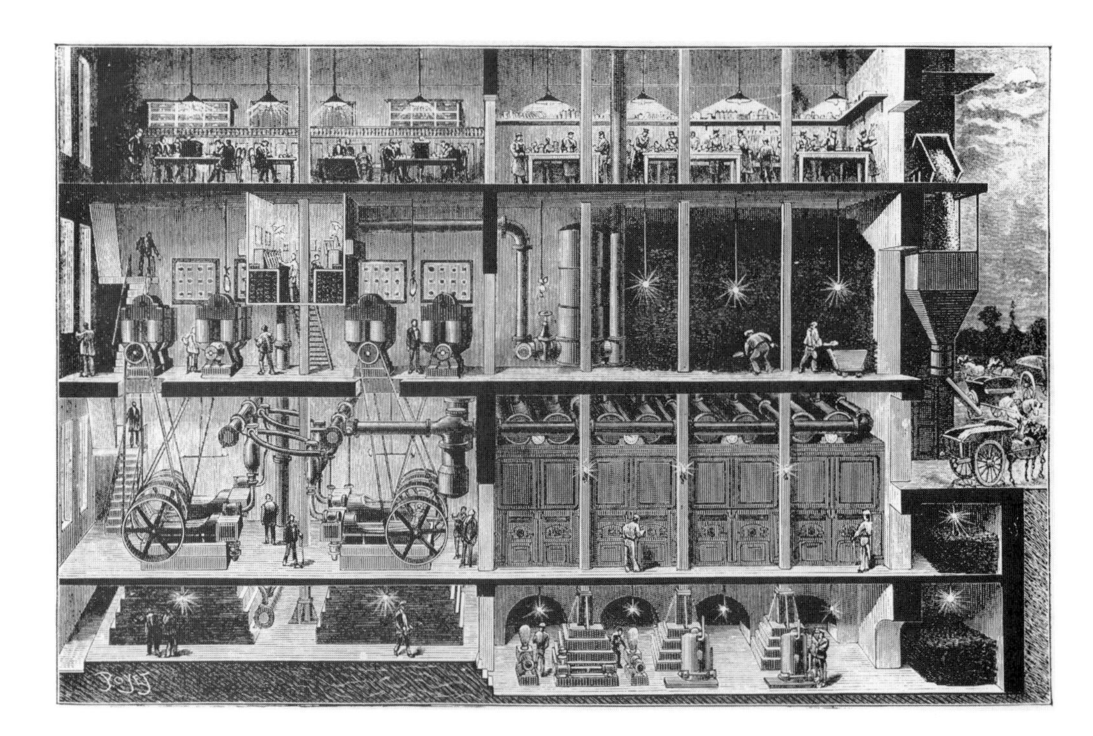

steam-driven dynamos, each with the capacity to light 1200 lamps. A completely new industry, the electrical industry, had been created from discoveries in the laboratory. 'Knowledge' had literally 'met with power'.

## 1.9 Science becomes a Profession

We may date the recognition that science is important to society by the invention of the word 'scientist'. It was invented in 1840 by the Reverend William Whewell. In his momumental survey of the sciences *The Philosophy of the Inductive Sciences,* 1840, Whewell wrote:

'We need very much a name to describe a cultivator of science in general. I should incline to call him a Scientist. Thus we might say that an Artist is a Musician, Painter or Poet, a Scientist is a Mathematician, Physicist or Naturalist.'

Whewell was writing at a time when, for the first time, there were sufficient people engaged in scientific work to make it a recognizable profession. He was also writing at a time when the 'cultivation of science in general' was becoming more difficult. It has been estimated that in 1840 there were no less than 500 scientific journals in print; science was growing fast and its fragmentation into specialities was well under way.

The first country to take the education of scientists and engineers seriously was France. One of the ideals which inspired the French Revolution, or at least its intellectual supporters, was the belief that the use of reason, and hence of science, would lead to a better society. This led to the foundation in 1794 of a major school of science and engineering, the École Polytechnique, and to the teaching of some science in the schools. Germany was not far behind; a most successful school of chemistry was set up at the University of Giessen in 1826 by the great German chemist Justus von Leibig, a man better known in Great Britain for his work on artificial fertilizers.

In Great Britain, despite its distinguished contribution to the scientific revolution, scientific education lagged behind Germany and France. The two major universities, at Oxford and Cambridge, were not really interested in science; they were deeply committed to the idea that a classical education is the essential foundation of culture. Moreover, they were run by men in Holy Orders and excluded dissenters who, by and large, tended to be the people most interested in science. Although science was regarded as a respectable hobby for the upper classes, it was considered unsuitable as a profession for a gentleman, let alone a lady; and so, in Great Britain, it had to enter the academic world through the back door. For blue-collar workers it entered through the Mechanic's Institutes which were founded in the 1820s; for the white-collar workers it entered through the colleges, like Owen's College in Manchester, which were being founded at that time in the major cities.

Thus, in the 19th century, science gained a foothold in the academic world and for the first time it was possible to undertake a recognized training. The German universities led the way; they proved to be outstandingly successful both in teaching and promoting research and for nearly 100 years they dominated academic science. In fact, until World War II most science students were expected to learn to read German. However, in due course most of the universities in Great Britain, USA and elsewhere, emulated the Germans. At Cambridge, for example, the 'Cavendish Laboratory for Teaching and Research in Physics' was built in 1871 and, by the end of the century, the number of science students in the universities of Great Britain had risen to a few hundred. Nowadays, some 80 years later, it is over 100,000.

Another important step towards the establishment of science as a recognized profession was the foundation of the specialized scientific societies. In the 18th century scientific societies were 'specialized' only in the sense that they belonged to a particular place; in the 19th century they became 'specialized' in the topics they discussed. In Great Britain, between 1800 and 1900, societies were founded for surgery, geology, astronomy, zoology, geography, entomology, botany, microscopy, pharmacy, chemistry and physics. They provided, so to speak, the 'social services of science'; they arranged meetings, conferences, publications, and fostered professional standards and recognition. They were, and still are, an essential part of the scientific community.

### 1.10 **Science grows fast**

In his presidential address to the British Association for the Advancement of Science in 1871 one of the greatest scientists of the 19th century, William Thomson (later Lord Kelvin) said: 'Scientific wealth tends to accumulate according to the law of compound interest.' Whether we consider that he was right or not depends on how we choose to define 'scientific wealth'. In simple terms he was obviously right; if we measure the wealth of science by its size, then it has certainly grown at compound interest for most of the 20th century. The number of people who, according to the statisticians, are engaged in 'scientific research and development' in western Europe and the USA increases by about 5 per cent per annum, which means that the total number doubles every 14 years. A simple calculation yields the astonishing result that about 90 per cent of all the scientists who have ever lived are alive to-day!

The topics of science have proliferated in much the same way. If you look in the current volume of the Dewey Decimal Classification for libraries you will find well over 1000 different classifications under 'Pure Science' alone. The growth of science is also illustrated by the numbers of scientific journals which are now published; it has grown from two in the year 1665 to well over 100,000 at the present

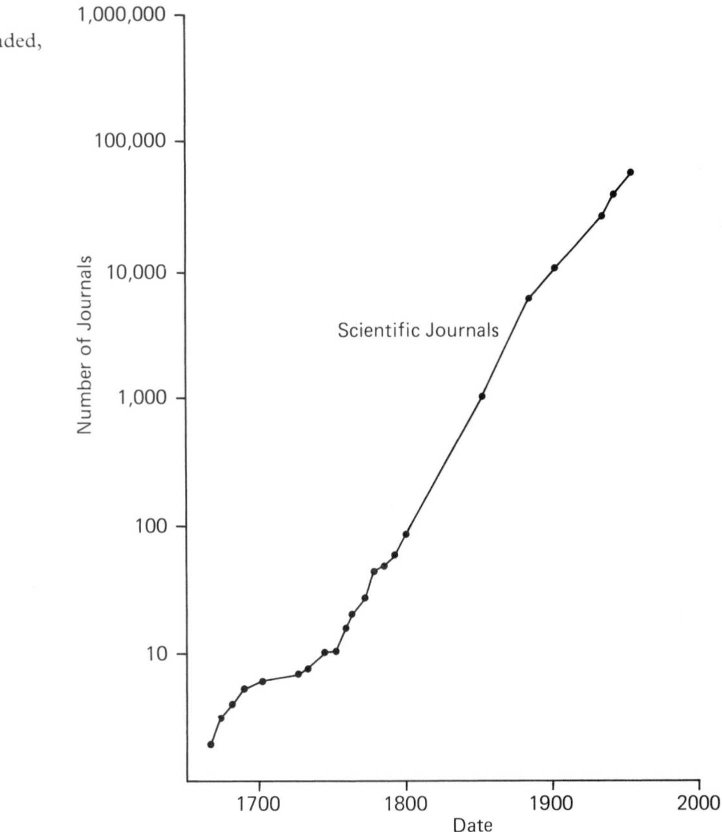


The total number of scientific journals founded, as a function of date.

time. Again, if we take the amount of money spent on scientific research and development we get much the same answer; for example, the amounts spent by the governments of western Europe and the USA were almost zero a century ago, but have now increased to a significant percentage of their Gross National Products (GNP). Broadly speaking, the 'size' of science, at least in the western world, has increased for many years 'according to the law of compound interest'.

Clearly this process cannot go on indefinitely. As Lord Bowden, the junior Minister of Science and Education in Great Britain, pointed out in 1965:

> 'For more than 200 years scientists everywhere were a significant minority of the population. In Britain to-day they outnumber the clergy and the officers of the armed forces. If the rate of progress which has been maintained ever since the time of Sir Isaac Newton were to continue for another 200 years, every man, woman and child on Earth would be a scientist, and so would every horse, cow, dog and mule....'

In fact the growth of science slowed down in the 1970's and in many cases has stopped; in the USA, for example, the total expenditure on R and D reached about 2.5 per cent of the GNP in 1970 and has remained roughly constant ever since.

If, however, we attempt to measure 'scientific wealth' by the significance of the published data, measured either in terms of advances in understanding or of practical value, and not simply by the amount of work done and money spent, then the answer is different and not so clear. It depends of course, upon how we choose our criteria. However, looking at the various attempts which have been made to assess the progress of science in terms of significance, there seems to be one point on which everyone agrees. The apparent rate of progress decreases as the criteria of significance are raised. Thus, at one extreme, if we take all the published data to have the same significance, then the 'wealth' of science has grown at a compound interest of 5 per cent, just like the other indices of 'size'. At the other extreme, if we select only data which, in the words of one study, 'convey really crucial insights that fundamentally enhance our picture of nature', then we find that significant scientific results have been produced at a steady rate which has not increased with time. Some such conclusion is only to be expected. As knowledge grows, it becomes harder to make progress; the problems are more complex, and the equipment more expensive, and so we must expect that the rate of significant discoveries will decrease with time. Perhaps that will not be such a bad thing; it will give people more time to get used to new discoveries.

## 1.11 **Science finds Patrons in Industry**

The foundation of the synthetic dye and electrical industries in the latter half of the 19th century was a big step forward in the development of science as a useful activity. Up to that time most scientists had been seen as trailing along behind industrial development, trying to rationalize and improve what had already been done. By contrast, these two new industries were based on novel discoveries which had been made by scientists working in research laboratories. Nowadays, this link between the research laboratory and industrial innovation, and the conviction that science pays dividends, are taken for granted in all except the oldest and most conservative industries. Indeed, many 20th century industries such as aviation, electronics, nuclear power and bio-technology, owe their foundation, if not their continued profitability, to scientific research.

A hundred years ago the total amount spent on research and development (R and D) by the 'industrial' nations was barely noticeable in their budgets. In the 20th century it has grown so fast that it is now a significant item. Broadly speaking this growth has followed the general growth of industrialization and trade but, in some countries, it has been accelerated more by the urgencies of war than by any far-sighted appreciation of the value of science.

In Great Britain, for example, the early stages of World War I – often called the 'Chemists' War' – exposed the scientific and technical shortcomings of British industries in the development and production of chemicals, explosives and dyestuffs, all of which were in short supply owing to the absence of imports from Germany. As a consequence, the British Government made a major effort to apply science to industry. It set up co-operative research associations for a wide range of industries, such as iron, glass, fuel, building, forestry, radio and so on. In practice, these associations served the smaller firms, because the larger firms, especially those concerned with the rapidly growing industries of chemicals, artificial fibres, electrical and radio and telephone equipment, were never content to depend on others for new ideas. At an early stage the very large firms such as General

Electric, Philips, Bell Telephone, and Du Pont established their own research departments which in due course grew to employ thousands of scientists and engineers.

World War II gave applied science an even greater boost; it has been called the 'Physicists' War'. The development of radar, the jet engine, penicillin, the large rocket, and operational research were all remarkable demonstrations of applied science, but they paled into insignificance by comparison with the atomic bomb. The explosion of the atomic

The development of large rockets in World War II greatly accelerated scientific studies in space. The Space Shuttle, Columbia, lifts off from the Kennedy Centre in Florida carrying the first Spacelab (28 November 1983).

bomb was a unique demonstration of the power of applied science for two reasons; it was incredibly powerful and, to most people, totally incomprehensible. It was not, like radar or the jet engine, a novel application of something reasonably familiar which most people felt they could understand if they tried, but was based on fundamental knowledge of the constitution of matter which only physicists could understand. Only a few years before, such knowledge would have been widely regarded as useless and arcane.

And so once again, the growth of research and particularly of development was stimulated by war. For example, in 1935 the total expenditure on R and D in the USA was only 0.1 per cent of the GNP; 15 years later, shortly after World War II, it had increased to 1 per cent. These wartime developments, and in particular the atomic bomb, compelled the attention of everybody to the power, for good or evil, of applied science. People asked why, if we can apply science so effectively in war, can't we make it work equally well in peace? Many people came to believe, as Francis Bacon had urged over 300 years before, that in the application of science to the needs of society lay the hope of a better world. Science was now widely recognized to be an important factor in industrial progress.

If we look at the expenditure on R and D by different groups in modern industry, we find, as we would expect, that it is greatest in the latest 'science-based' industries. For example, in firms which make motor-vehicles, scientific instruments and plastics it is typically between 5 and 10 per cent of their turnover, while in even more recent industries, such as computers or bio-engineering, it is even higher. On the other hand in the older industries, such as iron and steel or construction, it is far lower, typically 1 per cent or even less.

The majority of industrial R and D is now done by very large firms. A modern research laboratory is not only very costly to install and to run, but also must be reasonably large if it is to be effective. The minimum workable size is in most cases beyond the resources of small firms. In the USA, for example, the 300 largest firms account for 92 per cent of the total expenditure on research. Roughly speaking, all firms with more than 5000 employees have some sort of research department, while 9 out of 10 firms with less than 500 employees have none. This does not mean, by the way, that all good new ideas are to be found in large firms. You have only to drive along Route 128 around Boston to see that this is not so. It is lined with small firms making advanced products, many of which are based on bright ideas which were originally produced in the research laboratories of universities such as MIT.

Another effect of the increasing cost and elaboration of scientific research is that it has concentrated industrial R and D into the more wealthy of the so-called 'developed countries'. Thus, if we adopt the conventional distinctions between developed and developing, we find that roughly 97 per cent of the world's R and D, measured by cost, is done in the developed countries. Furthermore, the developed countries have about 10 times more engineers and scientists per head of population than the developing countries. One obvious and undesirable effect of this imbalance is that the majority of the world's R and D is influenced more by the needs of the more wealthy developed countries than by the more urgent needs of the developing countries.

### 1.12 **Science finds Patrons in Government**

Governments have always supported odd bits of scientific work which were urgent or happened to interest someone in a position of power; in earlier days a popular subject was the development of the calendar and, later on, of marine navigation. If we go back several centuries before Christ we find that the ancient Babylonian calendar was so complicated that it had to be regulated by mathematicians and astronomers who, presumably, were supported at public expense. Again, both Julius Caesar in 45 BC and Pope Gregory XIII in 1582 engaged astronomers to reform the calendar – quite a difficult scientific job! To take an example from marine navigation, the Portuguese, under the guidance of Prince Henry the Navigator, established in 1420 a considerable observatory at Sagres which was charged with research into navigation. It was quite a large 'research department' complete with cartographers, astronomers and instrument makers. About 250 years later, and on a much smaller scale, the British Government set up the Royal Observatory at Greenwich with much the same objective.

But most of these examples represent ad hoc responses to some limited technical problem, and it is not until the 20th century that we find governments beginning to recognize their responsibility to support systematic scientific research into fields which were unattractive to industry, such as public health and agriculture, and to realize the importance of science to war. We may date, for example, the first systematic commitment of the British Government to medical research by the National Insurance Act of 1911 which provided that 'one penny per insured person' should be set aside for research. The recognition of the Government's wider responsibilities for research was accelerated, as we have already noted, by the urgencies of World War I. The Department of Scientific and Industrial Research was set up in 1916 and was soon followed by Councils for Medical and Agricultural Research. Through these Councils the Government itself became involved in research into all sorts of topics, Food, Fisheries, Industrial Fatigue, Geology, Roads, Standards of Measurement, and so on. By 1939 the Department of Scientific and Industrial Research had a permanent staff of 2000 and was spending about £1m each year. Although it did not yet command as large a fraction of the national budget as it did later, government-sponsored R and D was firmly established in the years between the two World Wars and, as a result, science began to have an impact on many aspects of society which had not previously been touched by industry.

An early example of government patronage of science. The Royal Observatory at Greenwich was established by Charles II in 1675.

Following World War II there was yet another rapid expansion of government-sponsored science. For example, in the fiscal year 1969-1970, when this post-war expansion had slowed down, the total expenditure on R and D by the British Government (£560m) had grown to be larger than the total expenditure on R and D by the whole of British Industry (£460m). A major reason for this great increase in government spending was the inordinate complexity and rapid obsolescence of modern armaments and hence the need for more military R and D. The huge cost of military science has proved to be a permanent feature of most national budgets. In 1970, 25 years after the war, no less than 42 per cent of the British Government's expenditure on R and D was devoted directly to military work. Alas this is still true to-day, and not only in Great Britain. On the average, the principal industrial countries, with the exception of Japan, devote about one half of all their government-sponsored R and D to military applications. Japan is an exception because military work is prohibited there by the peace terms of World War II. Not all this great expansion of government-sponsored science in the present century has been due to military work; part of it has been due to the increased involvement of modern governments in matters of public concern such as education, health care, the purity of food, the protection of the environment, and so on, or in other words to the growth of the welfare state. For example, in Great Britain a major item in the R and D budget of the Government is a grant to the five Research Councils which represent the Government's involvement in Medicine, Agriculture, the Environment, and the Physical and Social Sciences. One of the important functions of these Councils is the support of long term basic research. Most industrial R and D is concerned with short-term economic objectives and, typically, is aimed no more than five years ahead. In practice this means that very little long-term basic research is done by industry and that most of the basic research in Great Britain is supported by the Government.

### 1.13 **Science is identified with its Applications**

Our widespread modern belief in progress was originally inspired, and is largely sustained, by the applications of science. By and large these applications have been so successful that, as we have seen in this chapter, the effort devoted to scientific R and D by the developed countries has grown in the last 100 years to absorb about 2 per cent of their GNP. Roughly one person in 300 of their populations is now classified as a scientist or engineer, and about 90 per cent of all their work is devoted to applied science. As a result, we have learned to see science-based technology as the most important face of science and to identify science almost wholly with its applications. We have followed Francis Bacon in equating the value of a scientific truth with its utility, but not, as he did, in identifying the betterment of mankind with the glory of God!

And yet there is another face of science which is concerned more with understanding the world than with transforming it. This other face has been rather in the shade since science became so useful. Let me, in the next chapter, try to throw some light on it.

# 2 Interpreting the World

'The philosophers have only *interpreted* the world in various ways, the real task is to change it.'
Karl Marx.

## 2.1 The Medieval Model of the World

In their efforts to interpret all that they know about the Earth and the Heavens people have always made models of the Universe. Sometimes they have pictured it as a 'cosmic egg', sometimes as a bowl carried on the back of a turtle. One of the most beautiful models is to be found in Ancient Egypt where the vault of the sky is formed by the goddess Nut arched over the recumbent body of the earth god Geb, touching the ground only with her toes and finger-tips; the Sun in this model was carried around the Earth in a boat and spent the hours of darkness in the perilous caverns of the underworld fighting off attacks by a serpent who lived in the depths of the celestial Nile and occasionally – during solar eclipses – succeeded in swallowing the Sun. Nearer our time, the last really successful attempt to fit everything that was known about the world into a coherent picture was the Medieval Model of the Universe, so vividly depicted by Dante in the 13th century. It is a convenient point at which to start this

Nut, the Sky goddess, above the recumbent body of Geb, the Earth god. Shu, the god of the Air, is holding up the solar boat.

chapter because it was the last world-view in which science played only a minor part; its destruction and replacement by a picture in which science played a major role, stands at the beginning of modern science.

The Model described by Dante, was based on the work of the theologians of the Christian Church in the 12th century who were trying to reconcile Christian theology with Greek science. Following Aristotle and Ptolemy the Earth is represented as a stationary globe at the centre of nine translucent

Medieval Universe presented by Dante in the *Divine Comedy*.

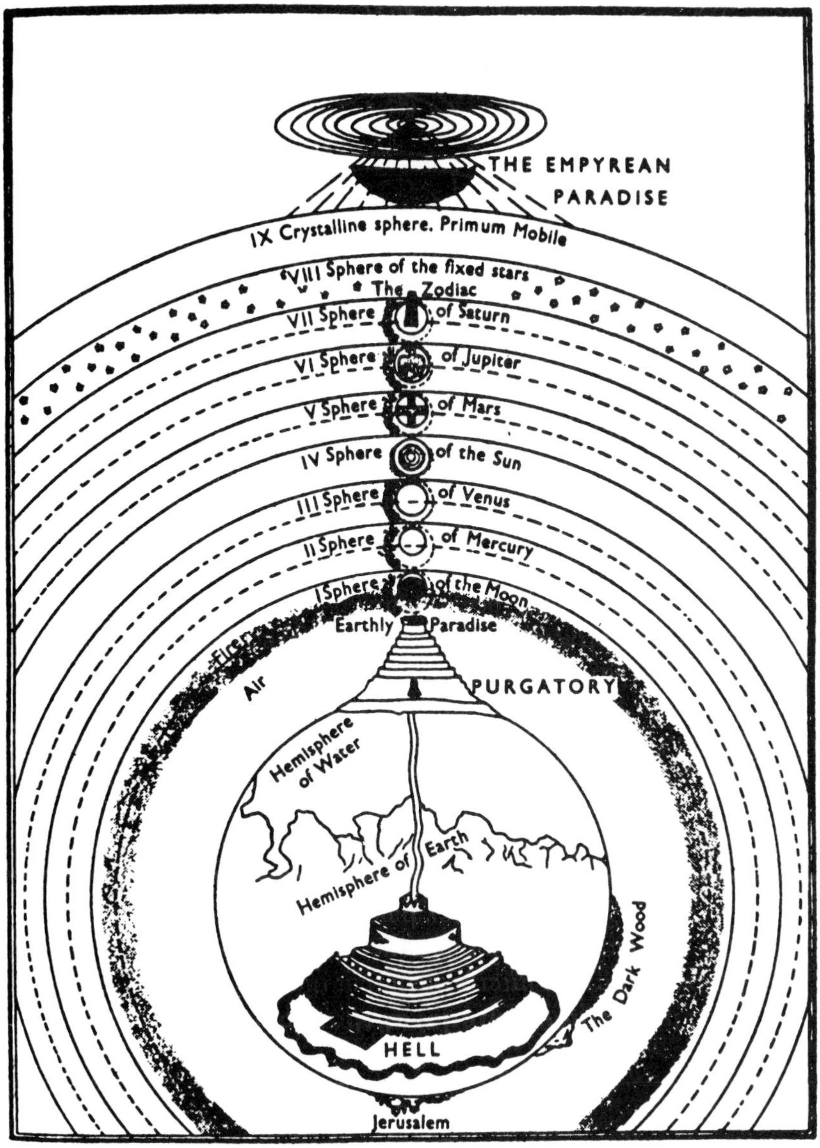

and revolving spheres. Eight of these spheres carry the Sun, Moon, five Planets and the fixed stars; the ninth, the *primum mobile*, drives the whole thing round. Outside the ninth sphere is the tenth Heaven, the eternal and infinite abode of God, which cannot be described in simple terms of space and time. To ask what is outside the tenth Heaven is like asking a modern cosmologist what is outside the expanding Universe; he will tell you that, by definition, the Universe contains all that *is*, and that therefore your question is meaningless.

A central feature of this Medieval Model is a sharp distinction between the nature of things terrestrial and celestial. Everything beneath the Moon is composed of the four elements of ordinary matter, earth, water, air and fire; everything above the Moon is made of a more perfect type of matter, a fifth element, aether. These two types of matter behave quite differently. Ordinary matter is always seeking to be in its proper place in the Universe; the most humble element, earth, seeks to be at rest in the lowest possible place at the centre of the Earth; water seeks to be above earth, air to be above water and fire to be above air. As a consequence, earth and water have a tendency, like apples on trees, to fall, and air and fire tend to rise. Horizontal motions on Earth, quite reasonably, are always the result of an applied force. In order to account for the fact that bodies, such as arrows and stones, continue to move without anything pushing them, the theory of *impetus* was elaborated in the 14th century. Impetus, so it was said, is implanted in a body when it is set in motion and acts as an internal force which is dissipated by resistance. A body, like a farm cart, only moves if it is pushed or pulled, and its speed is proportional to the applied force and to the resistance it encounters; when the force is removed the body, again like a farm cart, stops.

If left undisturbed, the four elements of basic matter would sort themselves out into four concentric spheres, like the spheres of Heaven. But in practice, they don't do this because they are constantly being disturbed at the boundary between ordinary and celestial matter; as the sphere of the Moon rotates, it churns up the outer layer of terrestrial matter which prevents ordinary matter from settling down. Thus, in the last analysis, all motions on Earth are due to the Heavens.

Celestial matter, unlike ordinary matter, is not subject to change or decay; furthermore, it obeys radically different laws of motion. Because it is already in its proper place in the scheme of things, it has no tendency to rise or to fall out of the sky; on the other hand, it is free to move sideways in perfect circles around the centre of the Earth. And so the nine celestial spheres, carrying everything we see in the sky, rotate about the Earth, each at its own speed. Each sphere is in the care of a Resident Intelligence, a sort of Angel, and is driven by the love of God or, in a later view, by 'impetus' imparted by God.

For several centuries this beautiful model answered most of the questions which people have always asked, and always will ask, about the origin, structure and purpose of the world, but in due course it was overtaken by the advance of science.

## 2.2 **The Medieval Model is destroyed – the Copernican revolution**

It was the astronomers and mathematicians of the 16th and 17th centuries who finally destroyed the Medieval Model. The first major attack was made in 1543 by a Canon of the Roman Church, Nicholas Copernicus. The work of Copernicus stands on the boundary between modern and medieval science. For instance, he accepted the medieval doctrine that celestial bodies must move in perfect circles at uniform speed, and he strongly resented what he regarded as the mathematical trickery which had been used in the later versions of the Ptolemaic system to improve their accuracy while still preserving the appearance of uniform circular motion; in particular he objected to the use of the equant, a geometrical device by which uniform angular motion took place about a point which was not at the centre of the circular motion. Paradoxically Copernicus undermined the Medieval Model in an effort to make it more consistent with the medieval principle that all celestial motions are necessarily circular. It seems, however, that his motives were not entirely mathematical. Apparently he was influenced by the Neoplatonic idea that the Sun is the visible form of God and must therefore occupy the most noble place at the centre of the Universe.

The idea that the Sun is at the centre of the Cosmos was certainly not new; it had been put forward several times before, notably by Aristarchos of Samos in the 3rd century BC. What Copernicus did was to show mathematically that such a model would work, and that it was, at least in principle, simpler than that of Ptolemy. As a result, his great book *De*

*Revolutionibus* was far too technical to be popular, and for over 50 years was regarded as little more than a moderately useful guide to the calculation of planetary positions. Nevertheless, his work survived because his model was seen by many astronomers to be attractively simple in conception, and because it was published at a time when astronomical evidence was starting to cast doubt on medieval physics.

The next really serious blow to the credibility of the Medieval Model was delivered by the great observational astronomer, Tycho Brahe. He recorded and published in a little book *De Stella Nova* a meticulously detailed account of a bright star which suddenly appeared in the constellation of Cassiopeia in 1572. For a short time it shone as brightly as Venus, and yet by the year 1574 had disappeared. Apparently something was wrong with the idea that celestial bodies never change. A few years later Tycho Brahe showed that the great comet of 1577 was at least three times as far away as the Moon, and was therefore far beyond the earthly regions of

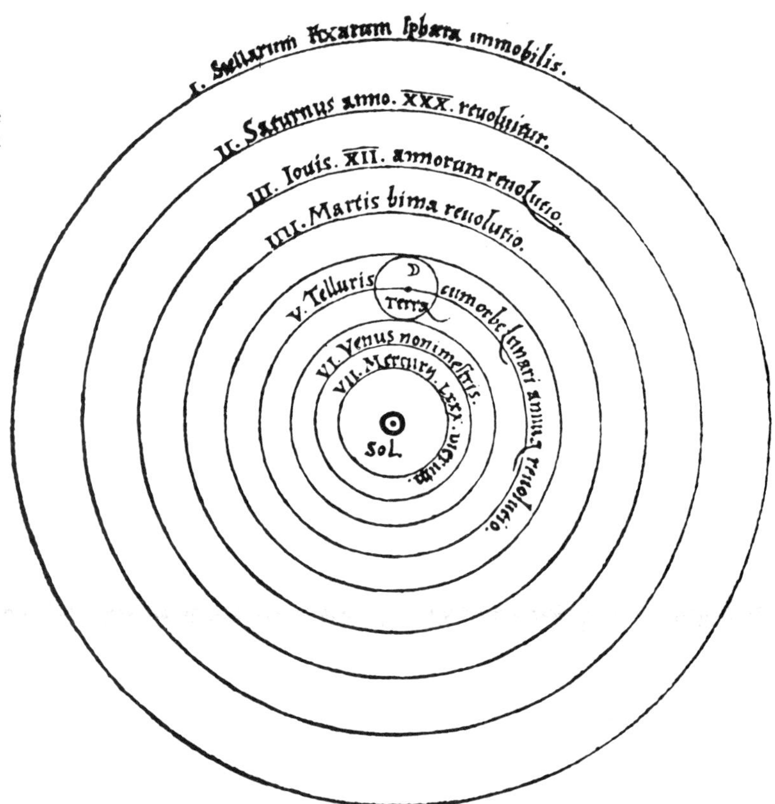


The revolutionary new model of the universe put forward by Copernicus in 1543. The Sun, not the Earth, is at the centre of the whole system and the Earth (circle V) is a planet of the Sun. From Nicholas Copernicus *De Revolutionibus Orbium Coelestium, Nuremberg*, 1543.

change and decay and well inside the supposedly unchanging Heavens. Quite apart from the obvious fact that the comet was evidence of change in the wrong place, there was the embarassing question of how it could travel through the solid translucent spheres.

The next blow was struck by a mathematician, Johannes Kepler, when he published his *Commentaries on the Motions of Mars* in 1609. By analysing the extraordinarily comprehensive and accurate observations of Mars bequeathed to him by Tycho Brahe, he showed that, contrary to orthodox belief, it is impossible to represent the motion of a planet by a circle or combination of circles because its orbit is an ellipse. Something was wrong with the idea that celestial bodies always move in perfect circles.

By the time Galileo turned his telescope on to the sky the Medieval Model was already in ruins, but that fact was not generally accepted. Galileo helped to complete and publicise its destruction. He showed anyone whom he could persuade to come and look through his telescope that there are mountains on the Moon and spots on the Sun which change from day to day, and that there are moons which circle about Jupiter and not about the centre of the Earth. Furthermore he published the arguments for the Copernican system in a book *The Two Chief World Systems*, 1632, which everybody could read because it was written in Italian. When, as a result, people began to take the Copernican system seriously and to believe that the Earth really does move, the Catholic Church had to act. The Church had invested too much of its authority in the Medieval Model, and its battle with the Protestants had made it so sensitive to heresy within its own ranks that it had become markedly anti-scientific. In an effort to stop the rot it suppressed the work of Copernicus in 1616 and put Galileo under arrest in 1633. But it was all too late; the Medieval Model was already beyond repair. Science came to a standstill in Italy, but elsewhere it was advancing and the time was ripe for a new world-view. There was not long to wait; Isaac Newton was born in 1642, the year that Galileo died.

## 2.3 Newton makes a new Model of the World

Having destroyed the Medieval Model the scientists of the 17th century faced the difficult problem of putting something better in its place. Admittedly Copernicus had shown them that, in principle, a simpler model should have the Sun in the centre, but in its fully developed form his system was almost as complicated as the later forms of the Ptolemaic model; it also failed to predict the positions of the Sun, Moon and planets with any greater accuracy. The basic reason for this failure was that Copernicus had retained the old idea that celestial bodies must move in perfect circles at uniform speed.

Before any new model could succeed there were three key questions about motion which had to be answered. Firstly, if the planets are not really carried round the Earth by transparent spheres driven by the love of God or by impetus, then what makes them move? Secondly, if they are not held in

The frontispiece of Galileo's *Dialogue Concerning the Two Chief Systems of the World, 1632.* From left to right are Aristotle, Ptolemy and Copernicus. This book publicized the arguments for the Copernican System and lead to Galileo's arrest and condemnation by the Roman Church.

place by transparent spheres or guided by Resident Intelligencies, then what holds them in their orbits? Thirdly, if, as Aristotle taught, a heavy body falls to Earth because it is seeking its proper place near the centre of the Universe, how are we to explain its behaviour if the centre of the Earth – and therefore the centre of the Universe – is moving?

In the course of the 17th century many scientists contributed essential pieces to the answers. Galileo, as we are inevitably reminded by the leaning tower of Pisa, experimented with accelerated motion; Descartes put forward the idea that a body will continue in a straight line unless it collides with another body; Robert Hooke, the curator of the Royal Society, suggested that there must be a force of attraction between all bodies which holds the planets in their orbits and causes things to fall to Earth. But no-one could show how powerful this force on the planets would have to be, nor how it must vary with distance, in order that they should move in Kepler's ellipses.

It was while seeking an answer to this problem that Edmond Halley, later Astronomer Royal, consulted the Lucasian Professor of Mathematics at Cambridge, Isaac Newton, and found to his astonishment, that Newton had already solved it but had mislaid the calculations! Halley persuaded Newton to write up his results in a book which Halley undertook to publish at his own expense. In this astonishing book, *The Mathematical Principles of Natural Philosophy*, 1687, commonly known as the *Principia*, Newton solved all the outstanding problems of motion, and then in Part III, called the *System of the World*, completed the Copernican Revolution by giving the world a new working model of the solar system. He showed, mathematically, how the motions of everything that we see in the sky – stars, Sun, Moon, planets and comets – can be explained and predicted by three simple laws of motion and one law of universal gravitation. He went on to show that all the motions which we see on Earth, everything from the rise and fall of the ocean tides to the behaviour of projectiles and falling apples, are governed by these same laws. The terrestrial and the celestial no longer obeyed different rules.

Newton's *System of the World* answered all the three key questions. The answer to the first question – what makes the planets move? – was given by Newton's first law of motion, which states that 'a body continues in a state of uniform motion in a straight line unless it is compelled to change that state by a force impressed on it'. Nowadays we accept this law without question, but it is contrary to common sense. Indeed our popular beliefs about motion have not changed since medieval times as much as we would like to think. Recent experiments in the USA have shown that most people believe that, when a stone is dropped by someone who is moving forward, it will hit the ground vertically below the place where it was dropped. They do not realize that the stone continues to move forward as it falls. Intuitively they think, as Aristotle had taught, that a body only moves in response to an applied force – as everyone knows, if you stop pushing a farm cart, it will stop. But as Newton showed, Aristotle had picked up the wrong end of the stick; a body only changes its motion in response to an applied force. Thus a planet, once it has been set in motion, goes on moving, quite simply because there is nothing to stop it, while friction stops the farm cart.

The answer to the other two questions was given by Newton's second law of motion (force mass × acceleration), and by his law of universal gravitation which tells us that the planets and the Moon are held in their orbits by the force of gravity; to account for this, and for the acceleration of falling bodies on Earth, Newton showed that there must be a force of attraction between any two bodies proportional to the product of their masses and inversely proportional to their distances apart. Furthermore, he proved mathematically that the force of attraction exerted by a large body, such as the Earth, acts as though all of its mass is concentrated at its centre; a falling apple really is attracted to the centre of the moving Earth.

In his *System of the World* Newton demonstrated the power of reason to explain what we see in the sky in terms of what we see on the Earth. He showed that the fall of an apple is governed by the same laws as the motion of the Moon and, by so doing, removed the ancient barrier between terrestrial and celestial physics. After Newton, science took on the job of explaining the whole visible Universe.

## 2.4 **The Mechanical Philosophy – the Cartesian revolution**

For most of the 17th century scientific thought was influenced, if not dominated, by the work of Descartes. In his *Discourse on Method,* 1637, published long before Newton's

*Principia*, Descartes tells us that there is but one kind of matter in the Universe, and that all the properties of matter which we perceive can be explained in terms of its division into parts and the motion of those parts. The only events which physics has to consider are the transfers of motion between particles, and the changes in the direction of their motion. The total quantity of motion in the Universe is constant and was put there in the beginning by God; all man can do is to alter its direction.

Some 50 years later, in 1687, Newton wrote in his *System of the World*;

'I am induced by many reasons to suspect that all the phenomena of nature may depend upon certain forces by which the particles of

The difficulty of reconciling Newton's ideas with common sense (a contemporary cartoon).

bodies, by some causes hitherto unknown, are either mutually impelled towards each other, and cohere in regular figures, or are repelled and recede from each other.... the whole programme of science is, from the phenomena of motions, to investigate the forces of nature and from these forces to demonstrate the other phenomena.'

Descartes and Newton were both expressing the same basic beliefs of a scientific world-view which, gathering strength in the 17th century, inspired the rapid rise of modern science in the next 200 years. I shall call these beliefs the 'Mechanical Philosophy', although that term is often restricted to the view, favoured by Descartes, that all motions are transmitted by physical contact. In that restricted sense Newton was not a true mechanical philosopher, because his *System of the World* made use of the force of gravity which, apparently, was transmitted from one body to another without any physical contact. Nevertheless his correspondence shows that he hoped that a mechanical explanation of gravity would eventually be found; in the meantime he was quite prepared to make use of the idea without such an explanation. As a contemporary of Newton, Bernard de Fontenelle, pointed out, Descartes started from what he clearly understood in order to find the causes of what he saw, whereas Newton started from what he saw in order to find out the causes whether clear or obscure. With that in mind we shall use the term 'Mechanical Philosophy' to embrace the ideas of both Newton and Descartes.

This new philosophy applied only to the world of matter which, following Descartes, was regarded as absolutely separate from spirit. The first article in its creed was that the world can be understood by the exercise of Reason; the second article, perhaps more a hope than a belief, was that this understanding can be expressed in terms of mechanical models with the help of mathematics. In this new 'scientific' world-view the world was seen as a succession of configurations of matter; physical events were no longer governed by human qualities and purposes, as people had believed in earlier times, but were determined by the mechanical interaction of inanimate objects obeying universal, mathematical, laws relating cause to effect. Planets were no longer moved by the love of God and falling objects did not aspire to reach their proper place in the scheme of things; to be sure, they still obeyed symbolic powers, but the new symbols corresponded to *measurable quantities* such as mass, force and velocity, the algebraic symbols of Newton's laws of motion. Robert Boyle, one of the leading scientists of the 17th century, put the whole thing rather crudely, but clearly, when he wrote:

'I look upon the phenomena of nature to be caused by the local motion of one part of matter hitting against another.'

After centuries of having to accept the authority of scholastic philosophy it must have been exciting to meet new, radical, ways of thinking about the world.

In the years which followed the publication of the *Principia* many branches of learning tried to apply the principles of the Mechanical Philosophy to their own problems. In the 18th century much of this work was unproductive, for example the problems of chemistry and physiology were not easy to analyse and quantify; but the mathematical astronomers were outstandingly successful, especially in France and Germany, and for most of that century they made celestial mechanics the senior science. They put Newton's laws of motion to the most stringent tests by tackling and solving a variety of extremely difficult questions concerning the motions of Jupiter, Saturn and the Moon. As we noted in §1.7, to calculate the motion of the Moon was a problem of considerable practical importance.

By the end of the 18th century the Mechanical Philosophy was firmly established and mathematical physics – eventually dynamics – was the leading model of human knowledge. So much so that when Napoleon Bonaparte asked Laplace why

The world seen as a mechanism. A late 17th-century orrery.

God was not mentioned in his monumental work *Mécanique Céleste, 1799*, Laplace replied: 'Sire, I have no need of that hypothesis.'

Let us look briefly at some of the successes of that philosophy in interpreting the world.

### 2.5 The Idea that Matter is made up of Atoms

For thousands of years people have wondered what would happen if they were to cut up a piece of matter, say a lump of lead, into smaller and smaller pieces. Would they eventually reach a limit when the pieces could no longer be divided any further? If so, what would those pieces be like? For most of recorded history there have been two contradictory answers to this question. One answer, given by Anaxagoras in the 5th century BC, was that matter is infinitely divisible; there is no such thing as a smallest possible piece. The other answer, given by Democritus at much the same time, was that all matter is made up of indivisible particles, atoms, which are so small that they cannot be seen. Democritus, whose work has been lost, is reported to have believed that these atoms are separate and can move freely in the void, bouncing off each other or interlocking to form composite bodies; furthermore, it is said that he believed that there are many different sorts of atoms and that homogeneous substances have atoms all of the same shape and size. Over 2000 years later people were still arguing about the same questions because, until the advances in chemistry at the end of the 18th century, there was little evidence one way or another.

To the mechanical philosophers of the 17th and 18th centuries these questions were fundamental to their attempts to explain the world in mechanical terms. Many of them believed in the 'corpuscular' theory of matter, but they couldn't prove it. Newton, for example, believed that:

> 'God in the beginning formed Matter in solid, massy, hard, impenetrable, moveable Particles, of such Sizes and Figures, and with such Properties, and in such Proportion to Space, as most conduced to the End for which he form'd them.'

Furthermore, he made the interesting suggestion (§2.4) that, to provide an explanation of some phenomena, it might be necessary to assume that these particles could attract and repel each other. But all this was little more than speculation, there was no solid evidence for these particles or atoms.

An essential step forward was the recognition that some

substances are elements which, unlike chemical compounds, cannot be created or destroyed in chemical reactions. This recognition was part of the fundamental advances in chemistry which took place in the 18th century and which were prompted by the question which has interested human beings ever since the discovery of fire – what happens when something burns? In the mid-18th century it was widely accepted that all bodies contain a substance called *phlogiston* and that, when they burn, this phlogiston escapes. Substances which burn easily, like wood, are full of phlogiston; substances which won't burn, like stone, contain none. There were, needless to say, objections to this theory; for example, it had been noticed that some substances gain weight when they burn, which is difficult to explain by the loss of phlogiston unless, as was suggested, phlogiston has negative weight or, as it was called, levity!

The first step towards understanding what really happens when something burns was to establish by experiment that all gases are not the same, and that there are different gases with different chemical properties. For example, in 1766 Henry Cavendish showed that there is a gas which he called 'inflammable air' (hydrogen) and which he made by dissolving various metals in acid. A few years later the work of Carl Scheele and Joseph Priestley established that ordinary air is a mixture of at least two different gases, one of which, 'dephlogisticated air' (oxygen), is involved in burning and respiration. It was Antoine Lavoisier who, on hearing of Priestley's work, showed that oxygen is solely responsible for combustion and, by combining what was already known about the elements with this discovery, gave us the first systematic account of the formation of chemical compounds from the elements. In his *Elementary Treatise of Chemistry*, 1789, Lavoisier gave a list of the known elements together with a systematic classification of their compounds, thereby laying the foundations of rational, quantitative chemistry. Fortunately, he completed this valuable work a few years before he fell foul of the revolutionary authorities in France and was guillotined in 1794.

The next step forward was made by John Dalton, the indefatigable secretary of the Manchester Literary and Philosophical Society. He recognized how the atomic theory of matter could be taken off the shelf of history and made to do useful work in explaining the laws of chemical combination which had been discovered so laboriously in the 18th century. In his *New System of Chemical Philosophy,* 1808, Dalton pointed out that the observed facts that chemical compounds contain fixed proportions by weight of their constituent elements, and that these elements are conserved in chemical reactions, could be explained by assuming that elements are composed of indivisible and indestructible atoms; all the atoms of any one element are identical, but the atoms of different elements differ in weight. Dalton's great contribution to atomic theory was to point to the significance of atomic weight.

Dalton visualized chemical compounds as being made up of 'compound atoms' composed of a smaller but definite number of the atoms of each constituent element. Later work, most notably by Avogadro (1871), put the study of these compound atoms on a quantitative basis and they came to be known as molecules.

The atomic theory of matter proved to be a powerful tool in the mechanical interpretation of nature. For example, the physicists of the 19th century succeeded in explaining the observed behaviour of gases in simple mechanical terms by relating their temperature, pressure, volume and other properties directly to the number, mass and motions of their

John Dalton's original list of the atomic weights of the elements.

ELEMENTS

| | W^t | | W^t |
|---|---|---|---|
| Hydrogen | 1 | Strontian | 46 |
| Azote | 5 | Barytes | 68 |
| Carbon | 54 | Iron | 50 |
| Oxygen | 7 | Zinc | 56 |
| Phosphorus | 9 | Copper | 56 |
| Sulphur | 13 | Lead | 90 |
| Magnesia | 20 | Silver | 190 |
| Lime | 24 | Gold | 190 |
| Soda | 28 | Platina | 190 |
| Potash | 42 | Mercury | 167 |

constituent atoms or molecules. On the other hand the chemists used the atomic theory to develop a quantitative theory of chemical reactions, but failed to explain these reactions in simple mechanical terms. It is interesting to note that although the atomic theory was widely used in the 19th century, there was still no irrefutable proof of the existence of atoms, and for roughly 100 years after Dalton it was possible to regard the atomic theory as a convenient, but not necessarily true, hypothesis.

It was not until the 20th century that developments in physics, not chemistry, showed beyond doubt that matter really is made up of atoms, although, as we shall see in §2.11, these 20th century atoms have turned out to be significantly different from those suggested by Dalton; for one thing, they

The atoms of a tungsten crystal magnified 10 million times by a field-emission microscope. (Courtesy of Professor T. T. Tsong of Pennsylvania State University.)

have proved to be divisible. In writing this I am irresistibly reminded of Descartes who in his *Discourse on Method,* 1637, argued that indivisible atoms cannot exist on the grounds that this would limit the omnipotence of God. God might perhaps have made particles so small that we ourselves cannot divide them, but He could not deprive Himself of the ability to divide them further, because that would limit His omnipotence. As we shall see later, we haven't yet reached the stage where we have to worry about that particular argument!

### 2.6 **The Idea that Living Matter is made up of Cells**

While 19th century science established the understanding that atoms and molecules are the basic building blocks of all matter, it also established the understanding that cells are the basic building blocks of living matter. Although, as we noted in §1.3, Robert Hooke used his primitive microscope in the 17th century to note that cork and some other vegetables have a cellular structure, it was not until substantial improvements had been made in the performance of microscopes (the development of the achromatic compound microscope) that people were able to study typical cells which have a size of about 10 microns (1 micron $= 10^{-6}$ metres). The first systematic and comprehensive observations of the cellular plants was made by Matthias Schlieden in 1838, and the extension of this work to all living matter, animals as well as plants, was put forward in a remarkably thorough study published by Theodor Schwann in 1839.

The scientific study of the cell, cytology, was now well and truly launched. The idea that the properties of living things might be explained in terms of their component cells was, like the atomic theory, very attractive to the scientists of the 19th century, and so the new science of cytology went ahead at a rate which was limited only by the power of the microscopes of that time. By the end of the century quite a lot was known about the cell; its major features (nucleus, cytoplasm, membrane, etc.) had been described and named, and the fact that cells differ widely in structure and function had been recognized. It had also been established that cells do not multiply by crystallizing from the cell fluid, as Schlieden had thought, but by cell-division. This fundamental observation was first made by the botanists, but was soon taken up by the zoologists and the process, called mitosis, was described in both plants and animals by Walther Flemming in 1880.

The discoveries made in the present century of the many remarkable functions of the cell, for example in the control of immunity to disease or in the control of growth in living creatures, have had to wait for the development of new instruments and for advances in biochemistry and molecular physics. It is only in recent years that the development of the phase-contrast microscope, the electron-microscope and the X-ray diffractometer have made it possible to study the very small components of the cell which are responsible for some of its most important functions and which are much too small to have been seen with an optical microscope.

## 2.7 **Understanding Heat and Energy**

Another major achievement of the Mechanical Philosophy in the 18th and 19th centuries was to answer an ancient question, what is heat? In the 18th century most people thought of heat as an invisible, weightless fluid which they called caloric – we still preserve this old name in our modern unit of heat, the calorie. It was believed that when two bodies at different temperatures come into contact the caloric flows from the hotter body into the colder and raises its temperature; the state of a body – solid, liquid or gas – depends upon just how much caloric it has absorbed. In the earlier part of the 18th century this theory was reasonably satisfactory because, for the most part, people were interested in mixing fluids at different temperatures and naturally they thought in terms of fluids. Later, however, when they became interested in steam engines, the theory was not so good. It failed to give a convincing explanation of the production of heat by friction or of the changes in temperature which were observed when a gas was compressed or expanded. Where, it was asked, did the caloric come from in these processes?

The first really serious attempt to test the caloric theory was made by Benjamin Thompson who was born in the USA, worked in Great Britain and Bavaria, and became a Count of the Holy Roman Empire (Count Rumford). While working on armaments for the Bavarian army he made a careful measurement of the heat evolved in boring a brass canon, and came to the conclusion that the heat that was generated in this way could not be limited by the amount of caloric in the brass, but must be produced by the mechanical work done. In the paper which he submitted to the Royal Society of London in 1798, he argued that his experiments had shown that heat cannot be a substance; he suggested that it is a mode of motion. Not surprisingly, we see that the important idea that *heat is a motion*, and not a fluid, came from mechanics and not from chemistry.

Count Rumford's conclusions were soon confirmed by Sir Humphrey Davy who performed an experiment in which two blocks of ice were melted by rubbing them together; even so, it was roughly 50 years before it was generally accepted that heat is not a fluid, but a form of disordered motion.

The understanding that heat is a form of motion and that mechanical work can be transformed into heat was put on a sound experimental basis by the work of James Prescott Joule in the 1840's. In a series of precise experiments he measured the 'mechanical equivalent of heat', and showed that in the exchange of heat and mechanical work energy is conserved. A new branch of science was born, thermodynamics; and if we are to believe C.P.Snow the ideas which underly this science should be more widely appreciated than they are. In his *Two Cultures,* 1959, Snow suggested that to be ignorant of the Second Law of Thermodynamics is culturally equivalent to never having read a work of Shakespeare!

The First Law of Thermodynamics was originally spelt out by a German doctor, Robert Mayer, in 1842. It tells us that although energy can take many forms, it cannot be destroyed – in other words it is the Law of the Conservation of Energy. The Second Law is just as fundamental but more interesting. It is based on the work of a remarkably talented young man, Sadi Carnot, an engineer in the French Army, whose original contributions to science received very little recognition in his lifetime. In his *Reflections on the Motive Power of Fire,* 1842, Carnot analysed the conversion of heat into mechanical work and pointed out that work can only be extracted from the heat in a body if, in the process, the heat flows to another body which is cooler; conversely, heat can only be transferred from a cool to a hot body by doing work. These conclusions represent one of the many ways in which the Second Law can be expressed.

Another, more highbrow form of the Second Law tells us that every physical and chemical process in nature takes place in such a way as to increase the total entropy of all the bodies taking part. Entropy is one of those concepts which many people find baffling. Broadly speaking, it measures the 'availability' or 'accessibility' of heat for conversion into work; the higher the entropy the lower the 'availability'. Interestingly it also measures the 'disorder' of a system, where 'disorder' is defined in terms of the statistical probability of finding the system in that state. Thus the Second Law tells us that in any natural process, if we take into account all the bodies involved, the disorder increases and the availability of heat energy for conversion into work decreases. This does not imply, as many people think, that the evolution of increasingly complex and ordered forms of life on Earth contradicts the Second Law of Thermodynamics. It means that when we take into account what is happening to all the bodies involved, including the generation of the energy which supports life on Earth by thermonuclear reactions in the Sun, then we shall find that the total disorder or entropy in the whole system is increasing. Clearly this rather depressing message is important outside the study of steam engines; cosmologists in particular, have interpreted it to mean that the whole Universe must eventually run down in what they call 'the heat death of the Universe'.

At first sight the Laws of Thermodynamics look rather too dull and too abstract to be useful; but that is deceptive. In fact, they were of great practical value in an age where muscular power was being replaced by machinery; they made it possible to relate and analyse a wide variety of previously unrelated physical processes, such as the conversion of the energy stored in coal into mechanical work by a steam engine, or the conversion of mechanical work into electrical energy. Indeed, the recognition of heat as a form of motion, and of energy as an indestructible and accountable feature of the world, were achievements of which the Mechanical Philosophy could be justifiably proud; but as we shall see in the next section, as a way of understanding Nature it wasn't always so successful.

## 2.8 **The Idea of the Aether – the Mechanical Philosophy fails**

You have only to look at the sky for a moment, or to pick up a pin with a magnet, to appreciate some of the difficulties of explaining the world in simple mechanical terms. How, for example, is the force of gravity transmitted across space and how does the light from the Sun travel to us across a vacuum? What actually pulls the pin towards the magnet? Descartes, one of the most enthusiastic mechanical philosophers of all time, had answers to all these questions. He believed that all space is filled with material and that there is no such thing as a vacuum. Like all true mechanical philosophers he did not believe in 'action at a distance' – things had to touch one another so that they could push and pull. In his view, the motions of the planets could be explained in simple mechanical terms by vortices in this all-pervading medium. Magnetic forces were transmitted by specially shaped particles which streamed through magnets. Light was transmitted across space as a simple mechanical force in the medium and travelled with an infinite velocity.

Newton shared Descartes' belief that the world could be interpreted mechanically, but he disagreed with him about vortices. He showed mathematically that vortices could not account for the motions of the planets and substituted the idea of universal gravitational attraction, namely that all bodies, such as the Earth and Moon, attract each other across space. However in putting forward this idea he did not, as we have already noted, give up his faith in the Mechanical Philosophy. Like Descartes, he believed that all space must be filled with an invisible medium, the aether, and conjectured that the force of gravity might perhaps be related to local variations in the density of this medium. As far as the transmission of light was concerned, the aether was not really needed; Newton thought of light as being transmitted by 'fiery particles' travelling at great speed.

As both Descartes and Newton had found, any thoroughgoing mechanical explanation of the physical world must have an aether. Quite apart from the task of transmitting the force of gravity it was needed to transmit electric and magnetic forces through space. By about 1850 it had been established that the corpuscular picture of light was wrong and that light behaves as a wave. Not unreasonably, it was argued that if light is a wave motion then there must be something in space to wave – there must be an aether. To support the transverse vibrations of light it was realised that this aether must be an elastic solid and yet it must not be so dense as to impede the motions of the planets. It was far from clear whether two or three different aethers were required to explain the propagation of light and also the transmission of electric and magnetic forces.

This question was clarified in 1861 when Maxwell developed a new and powerful idea – the idea of a *field*.

Following the experimental work of Faraday, in which the relations between electricity and magnetism were explained by 'lines of force', Maxwell developed the concept of electric and magnetic fields which could pervade a medium and transmit a static force or a travelling wave. Such fields did not involve any simple mechanical motions or linkages, but might be thought of as stresses or strains in the aether. Maxwell showed that the theoretical velocity of these waves was identical with the measured velocity of light, and it was soon appreciated that light, radiant heat and radio-waves are all the same sort of waves – electro-magnetic waves. Maxwell had shown that only one aether was required to transmit light, electricity and magnetism.

Although Maxwell showed more clearly what the aether had to do, he did *not* solve the problem of describing its mechanical properties. In fact, a prodigious effort to answer that question was made in the latter half of the 19th century, and some incredibly complicated models were put forward involving spheres, caps, bars and even flywheels. In retrospect these models look rather comical and remind us of the mechanical complications of the later models of the Ptolemaic Universe. However, none of these elaborate models was wholly successful and no mechanical explanation of how the aether could transmit the force of gravity, or of electromagnetism, was ever found.

Indeed, by the end of the century, it was recognized by many physicists that all the phenomena of nature could not be explained, as Robert Boyle had hoped (§2.4), 'by the local motion of one part of matter hitting against another'; furthermore it was now recognized that such explanations were not essential to the progress of science.

## 2.9 The Idea of Evolution – the Darwinian revolution

We have seen that in the 16th and 17th centuries there were two major advances in the physical sciences, the 'Copernican' and 'Cartesian' revolutions. In the 19th century there was a third major advance, this time in our understanding of the Life Sciences. The Theory of Evolution gave to the scientific world-view a new dimension, a sense of history through a realistic time-scale. Newton's *System of the World* had no historical dimension. Its celestial machinery was completely explained by the science of dynamics without appeal to the spiritual domain, but that was not true of its history. As far as

Newton was concerned, everything had been created by God as described in the Bible, about 6000 years ago, and at that time all the infinite variety of the world, animals, plants, minerals and stars, had been fixed for all time. There was no room for change or growth or 'progress'. The study of fossils and rocks in the 18th and early 19th century suggested that something was seriously wrong with Biblical history. In particular, the fact that the fossilized remains of different species of animals and plants were found in different strata of rocks, pointed to the conclusion that living things, like rocks, had evolved over periods of time unimaginably long compared with the Biblical story of creation. Charles Darwin and Alfred Wallace confirmed this conclusion. They showed that the immense variety of animals and plants, fossilized and living, could be seen as a sequence which had evolved over a period comparable with the geological time-scale. Furthermore, they put forward the brilliant idea of 'natural selection' by which the species which survive are those which are best adapted to their environment. Darwin's book *The Origin of Species*, 1859, spelt out the theory of evolution for everyone to read. It proved to be the most controversial and widely discussed book in the history of science.

An important shortcoming of Darwin's work was that he gave no reason why new variations were always arising in a species. Many people pointed this out, among them Bishop Wilberforce. The Bishop is usually remembered for asking that great champion of evolution, Thomas Henry Huxley, in the course of a debate in Oxford in 1860, whether it was 'through his grandfather or his grandmother that he claimed descent from a monkey?'. But Wilberforce is commonly underrated; he had something more important, if less amusing, to ask. He wanted to be shown that:

'there is actively at work in nature, co-ordinate with the law of competition and with the existence of such favourable variations, a power of accumulating such favourable variations through successive descents.'

At the time of the Oxford debate no one could demonstrate such a power and they didn't know that a Moravian monk, Gregor Mendel, was actually engaged in doing so.

Mendel was cross-breeding peas in the garden of a monastery at Brno. He chose seven different characteristics of one species of pea and showed how they were inherited to produce different varieties. From his results he derived the fundamental ideas of dominant and recessive characteristics and worked out the mathematics of their inheritance in successive generations. He published his results in 1865 in the *Proceedings of the Brno Natural History Society* where, not surprisingly, they stayed almost unknown for 35 years. When they were rediscovered in 1900, they gave to the theory of evolution what it badly needed, some idea of how heredity actually works.

## 2.10 **Fin du Siècle**

Science in the 19th century was supremely self-confident. The influence of the Theory of Evolution on what we now call the Life Sciences was comparable with that of Newton's work on astronomy; for the first time it was possible to make sense

*The Lion of the Season.* Alarmed Flunkey: 'Mr G-G-G-O-O-O-RILLA!' Punch's comment on Evolution (May 1861).

of the bewildering variety of plants and animals. But its influence was not confined to the Life Sciences; like the earlier work of Newton it spread into all branches of science and into the mainstream of contemporary thought. It encouraged people who studied topics such as ideas, customs, organizations, societies, and so on, to think in terms of adaptation and change; Karl Marx, for instance, acknowledged his debt to Darwin's *'Origin of Species'* in the development of his ideas about the role of class struggle in history. Broadly speaking, the idea of evolution reinforced the modern belief in the possibility of Progress which we have inherited from the 18th century, and greatly strengthened the basic belief of the scientific world-view – that the world can be understood by the exercise of Reason.

It was the hey-day of the Mechanical Philosophy; advances in chemistry and physics had demonstrated its great power as a way of interpreting nature. Indeed, the structure of what we now call 'classical physics' with its grand generalizations about dynamics, thermodynamics, and electromagnetism, was so impressive that many people thought that there was not much more to be done. America's first Nobel Laureate, Albert Michelson, put this clearly when he wrote in 1902:

'the more important fundamental laws and facts of physical science have all been discovered and these are now so firmly established that the possibility of their ever being supplanted in consequence of new discoveries is remote.'

It was soon to be proved that he was profoundly wrong.

## 2.11 **Science in the 20th century – exploring the very large and the very small**

Science and technology advance hand in hand; nothing illustrates this better than the progress of physics and astronomy in the present century. At one extreme our knowledge of the very large, the Universe, has been greatly expanded by the construction of larger and larger radio and optical telescopes and by the development of space observatories; they, in their turn, have been made possible by developments in low-expansion glass, electronics, rocket propulsion, electronic computing, and so on. At the other extreme our knowledge of the very small, the atomic and subatomic world, has been made possible by the construction of larger and larger accelerators, electron microscopes, and so on; they, in their turn, have been made possible by a whole

Three hundred years of progress in reflecting optical telescope: (*Below*) A model of the reflecting telescope made by Isaac Newton in 1668. The mirror has a diameter of 2 inches. (*Right*) The UK Schmidt reflecting telescope at Siding Spring Observatory, Australia. The mirror has a diameter of 48 inches. (*Bottom right*) The Hubble space telescope due to be launched in 1986. The mirror has a diameter of 95 inches.

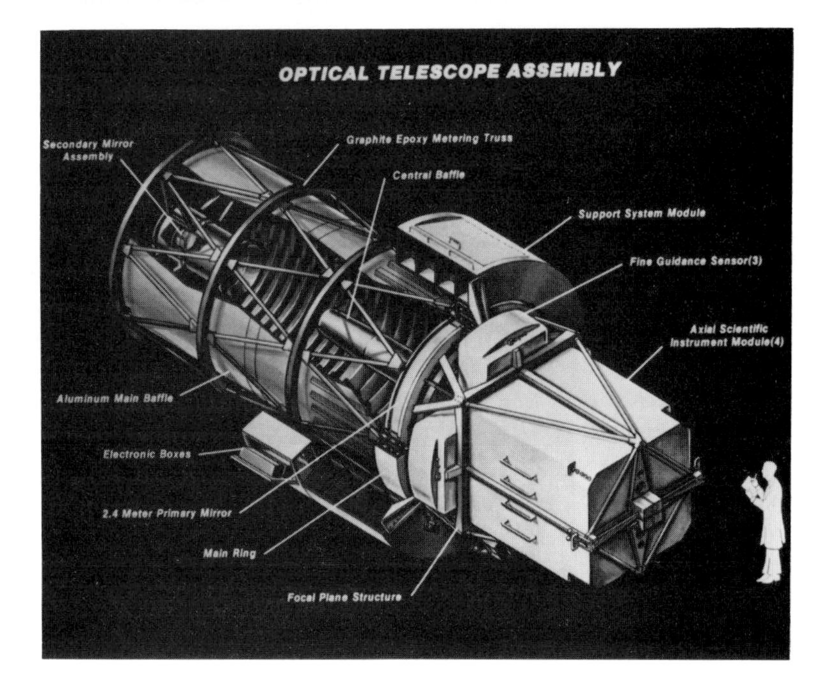

host of technical advances, such as the construction of very powerful magnets.

There have, of course, been many other significant advances in 20th century science, such as in our understanding of plate tectonics and molecular biology, but in what follows I have chosen to discuss some new ideas taken from atomic physics and astronomy because they illustrate two important changes in the scientific world-view in the present century. These changes are, firstly, the recognition of the limitations of classical science (the Mechanical Philosophy) as a way of understanding the world and, secondly, the recognition of the unimaginable size and age of the Universe. Advances in both these directions have brought with them some completely new ideas and have warned scientists that the physical world is much stranger than anyone could ever have imagined.

### 2.12 **Exploring the very small – the realm of the atom**

Let us look first at the exploration of the very small. We will start in a physics laboratory by looking at a phenomenon which, although discovered in the early 18th century, became one of the showpieces of late 19th century physics. We shall watch how an electric spark, passing between two metal electrodes in a glass container, changes from a thin spidery column into a beautiful glow as the air is pumped out. To do this experiment properly one must have a source of high voltage and a good vacuum pump. Both of these things were available in the laboratories of the late 19th century and it was the study of these beautiful, though apparently useless phenomena, which led to the important discoveries of the electron, X-rays and, indirectly, of radioactivity in the period 1895-1900.

The first discovery was dramatic and, as so often happens in science, it was accidental. Konrad Röntgen was observing an electric discharge in a vacuum tube in his laboratory in Germany when he discovered to his surprise that it emitted rays which penetrated the light-proof wrapping of a photographic plate. He called these mysterious rays 'X-rays', and for a short time they were one of the wonders of the world.

Röntgen's discovery engaged the attention of almost every practising physicist in the world and it was not long before two more discoveries were made. Only a few months later Henri Becquerel in France was trying to find out whether Röntgen's mysterious rays were associated with the phenomenon of phosphorescence. He chose to test uranium nitrate and made the surprising discovery that it, too, apparently gave off penetrating rays; in fact these rays proved not to be X-rays but fast particles called alpha-particles. Becquerel had discovered a completely new and most important phenomenon, natural radioactivity.

Röntgen's work on X-rays was also an important factor in another remarkable discovery, the discovery of the electron by J.J.Thomson in England in 1897. Thomson showed that the current in an electric discharge in vacuum is carried by minute charged particles (electrons); furthermore he showed, using Röntgen's X-rays that these electrons are one of the basic constituents of matter.

Thus at the start of the 20th century it was clear that the model of the atom which had served so well in the 19th century – in which the atom was pictured as a solid hard particle like a billiard ball – was too simple; it could not explain natural radioactivity, nor could it explain the electron. Physicists started on the long search – some would say the endless search – to understand the structure of matter.

One of the first things to do was to try to look inside the

The discoverer of the electron. J. J. Thomson giving a lecture-demonstration in the Cavendish Laboratory (Cambridge, England).

atom, and the obvious technique was to shoot things at it and see what happened. This was done by Ernest Rutherford who bombarded the atoms in a thin metal foil with the particles emitted by a natural source of radioactivity (radium) and measured where the particles went. Most of the particles travelled straight through the foil, but some bounced back; evidently they had hit something small and massive inside the atom, something very much heavier than the electron. Rutherford had discovered the atomic nucleus, and in 1911 he announced a model of the atom in which negatively

The discoverer of the atomic nucleus. Lord Rutherford (right) talking to J. A. Ratcliffe in the Cavendish Laboratory (Cambridge, England).

charged electrons circulated around a central nucleus, like planets going around a central Sun. It was recognized that the atoms of different elements have different numbers of positively charged particles (protons) in their nuclei and therefore have different numbers of orbiting electrons. At last it was possible to explain Dalton's conclusion that the atoms of various elements differ in size and weight, and to begin to understand why they have different chemical properties.

In those early days of atomic physics things were attractively simple; there were only three 'fundamental particles', the proton, the electron and the photon. The proton and electron were the basic building blocks of the atom and the photon was the basic building block of light. It looked for a time as though the ancient search for a simplicity underlying the complexity of matter had reached its goal. But the picture didn't last for long. The next stage was to try to look inside the nucleus and, once more, the obvious technique was to shoot things at it in the hope of breaking it open. This was accomplished by Rutherford in 1919 when he showed that the nucleus of the nitrogen atom could be broken open by a direct hit from an alpha-particle. It was now clear that to take a good look inside the atomic nucleus a new tool was required; what was needed was a source of fast particles with greater energy than those emitted by natural radioactivity. The first artificial particle accelerator was built by Cockroft and Walton in Rutherford's laboratory in 1931 and was used to break open the nuclei of some of the lighter elements. It accelerated protons to an energy of about 200 Kev ($2 \times 10^5$ electron volts) and cost about £1000 to build.

The race to unlock the secrets of the nucleus started. Accelerators were built in many countries and, ever since, the nucleus and its components have been bombarded with particles of higher and higher energy. In fact the energy of the particles has increased by a factor of ten times about every seven years since the first accelerator was built in 1931; the cost has increased in much the same way, but not quite so fast. As one recent example the accelerator at Fermilab near Chicago in 1975 can produce a beam of particles with an energy of about 500 Gev or $5 \times 10^{11}$ electron volts, which is about 2 million times greater than the original accelerator made by Cockroft and Walton. It is built in the form of a ring with a diameter of about 2 km and cost several hundred million dollars to build. In Europe the accelerator HERA at

Hamburg can collide 30 Gev electrons with 850 Gev protons and the Super Proton Synchrotron at CERN (Geneva) can collide protons with anti-protons. And that is by no means the end of the story, even larger machines are planned.

As a result of all this effort the number of particles which have been discovered since Bohr first put forward his model of the atom in 1913 has risen from 2 to about 200, and our picture of the structure of matter is now far more confusing than it was. However there have been strenuous efforts to simplify this picture. A theory put forward in 1963 proposed that the majority of the known particles, including the proton and the neutron, are made up of different combinations of a small family of more fundamental particles called quarks, a name borrowed from *Finnegan's Wake*, 'Three quarks for Muster Mark'.

Only three different kinds of quark, 'up', 'down' and 'strange', were originally envisaged, and for a short time it looked as though we had arrived at an attractively simple picture. Nowadays things are not quite so simple; at the time

The tunnel of the 7-kilometre circumference Super Proton Synchrotron at CERN (Geneva) where protons can be accelerated to an energy of 450 Gev. The SPS has been used to store proton and anti-proton beams circulating in opposite directions at the highest energies ever achieved under controlled conditions.

of writing there are 18 different kinds of quark and 18 anti-quarks! So far, the quarks are largely hypothetical, and there are theoretical reasons for believing that it may be impossible to detect an isolated quark. One can't help wondering how long it will be before the quark, like one of those Russian dolls, is split up into even smaller parts. The problem nowadays is no longer to break open the atom or the nucleus but to break open the components of the nucleus; that requires enormous energy and is the reason why such colossal particle accelerators are being built.

This story of the search for the secrets of the architecture of matter illustrates rather well many of the features of modern science. It starts, like so many other branches of science, with an accidental discovery and then shows how the construction of new tools, in this case accelerators, brings us entirely new knowledge of the world, knowledge which is so strange that it could never have been foreseen. It shows how our knowledge of the real world is limited by the tools which are available at the time, and goes on to show how each step forward in science depends on the interplay of imaginative theory (e.g. quarks) and experiment; furthermore it illustrates how each discovery in an active science raises more questions than it answers. Looked at more closely it shows how specialized and esoteric much of modern science has become; most scientists, even most physicists, find modern particle physics so complicated, so mathematical and so full of jargon that, even to them, it is largely incomprehensible.

Finally the development of particle physics illustrates how the character of much of present-day scientific work has changed. At the beginning of the century, work on the structure of matter was carried out by a few people in universities using equipment which the university could afford and which they themselves could build and operate: Rutherford's original equipment for splitting the nitrogen atom cost very little and could be held in the hand. The work is now carried out by large teams in national and international laboratories with equipment which costs millions of dollars and takes years to build. This exploration of the structure of matter has been one of the major achievements of the present century, and has brought results of immense importance to almost all branches of science and, of course, to industry as well. Biology and our understanding of genetics, to take but one example, has been revolutionized by advances in our understanding of the structure of molecules. But that is not all; the exploration of the behaviour of matter on an atomic scale has unearthed some completely new ideas which are not merely extensions of classical science; they are alien to it.

### 2.13 **The Ideas of the Quantum Theory – a revolution in scientific thought**

Of all the insights into nature which science has given us since Newton's *System of the World*, the ideas of the quantum theory are the most revolutionary. The quantum theory is concerned with the behaviour of matter in the atomic and sub-atomic world and, seen in historical perspective, marks a second phase in the history of modern science – the first phase being the development of the Mechanical Philosophy which is, broadly speaking, organized common-sense; but the quantum theory is organized uncommon-sense.

The development of the quantum theory was prompted by the failure of 19th century physics to explain the way in which the heat and light from a hot body vary with wavelength. The classical theory of radiation agreed with experimental results at long wavelengths, but was absurdly wrong at short wavelengths; in fact, it predicted that the energy radiated by a hot body would increase indefinitely at very short wavelengths. For that reason, the whole difficulty came to be known as the 'ultra-violet catastrophe'.

In 1900 this catastrophe was averted by Max Planck in an interesting way. First of all he found by trial and error that quite minor alterations to the classical formula for radiation made it agree with experiment, and then he used this modified formula to discover how hot bodies actually radiate. The classical theory of radiation assumed that the total energy of a body is infinitely divisible among all the possible ways in which it can absorb and radiate energy. To his surprise and dismay, Planck's new formula showed that this assumption is wrong; the energy radiated by a body does not come to us as a smooth and continuous flow, it comes in small packets called quanta.

The next step was taken in 1905 by Albert Einstein who showed that this same idea, that energy comes in small packets, explains the behaviour of radiation when it is detected as well as when it is generated. It was already known that when light falls on a metal surface it ejects electrons, but this phenomenon could not be explained by classical theory.

Einstein showed how these difficulties could be overcome if the beam of light was regarded as a stream of localized quanta of energy, later to be known as photons.

The next step was taken in 1913 when Niels Bohr showed that the concept of a quantum could also be used to understand another problem which could not be explained by classical physics, the stability of the atom. According to classical theory the electron could not continue to orbit the nucleus of the atom because it would radiate away all its energy. In Bohr's model of the atom, the electrons circled the central nucleus in one or more of a series of stable orbits in which their angular momenta and total energy were a fixed number of quanta. If an electron jumped from one orbit to another, due perhaps to a collision with another atom, it emitted or absorbed a quantum of energy corresponding to the difference in energy between the two orbits.

In terms of classical physics Bohr's model of the atom was nonsense, but it worked. Not only did it offer some explanation of the stability of the atom, but it also explained in detail some other puzzling phenomena, such as the spectral lines emitted by different elements.

Physicists were now obliged to recognize that, on an atomic scale, the behaviour of matter is not continuous, it jumps, and in the 20 years which followed the publication of Bohr's model of the atom a whole new branch of theoretical physics was developed to deal with this strange behaviour. This was the quantum theory; its development is one of the major intellectual achievements of the present century. We shall now look briefly at some of the new ideas which it has generated.

## 2.14 **The Idea of Complementarity – the observer enters the picture**

One of the great surprises of the present century has been to learn from physics how limited our common-sense ideas about reality are. This has been made especially clear by the application of the quantum theory to light. Consider, for a moment, the well-known experiment where a beam of light falls on an opaque screen in which there are two very narrow slits which are close together. If we look at the light which emerges through those slits on to another screen we see that, where the pools of light from the two slits overlap, there are alternate bright and dark bands. Since these bands, or so-called 'fringes', were first discovered by Thomas Young early in the 19th century, it has been realized that they show, beyond doubt, that light is a wave motion; the alternate bright and dark bands correspond to places where the light waves from the two holes alternately reinforce and cancel each other. On the other hand, if we now allow the beam of light to fall on a photo-electric detector where it ejects electrons, it behaves, again beyond doubt, as though it was a stream of discrete particles or photons. Clearly we must accept that in the first experiment light behaves as a wave, and in the second as a particle.

Let us now do an experiment in which light behaves both as a particle and as a wave. Let us repeat our first experiment with a very, very faint source of light and, instead of looking at the interference fringes, register the arrival of single photons at the second screen by means of a photo-electric detector. If we do this experiment for long enough to register many hundreds of photons and mark on the screen the positions of their arrival, we shall find that the pattern reproduces exactly the original pattern of light and dark bands which we saw on the screen when the light was bright and which told us that light is a wave. Many photons arrive in the bright bands and few in the dark bands. Somehow or other the photons seem to know where to land on the screen in the same pattern as if they were waves! To analyse this experiment theoretically physi-

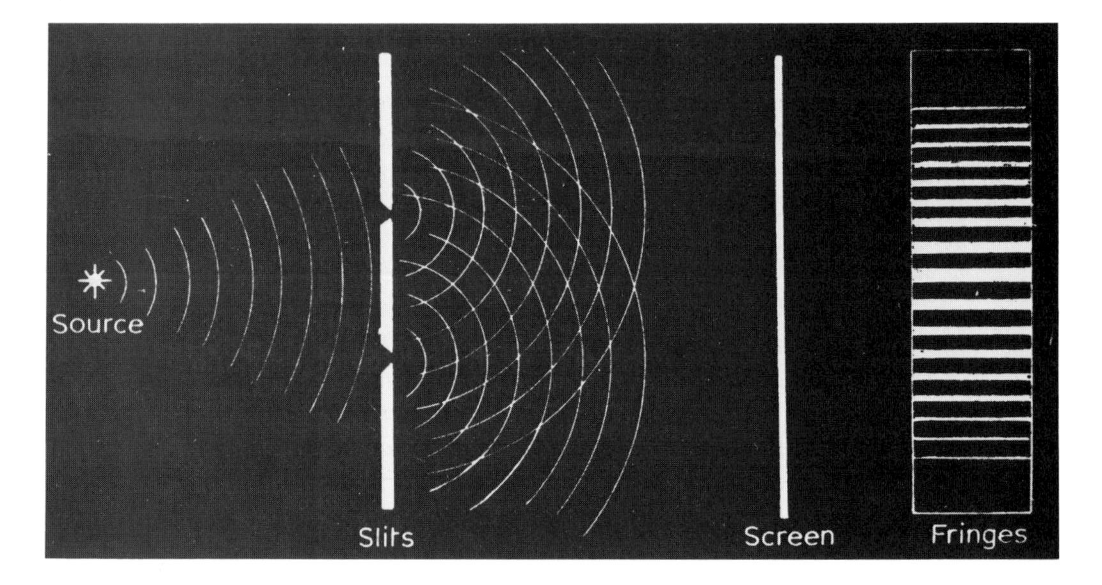


Young's fringes. Light passing through two closely spaced slits 'spreads out' beyond them to create a pattern of interference 'fringes' on a screen.

cists first treat the light as a wave, and calculate by classical wave theory how its intensity varies over the screen; they then use the calculated intensity at any point on the screen as a measure of the *probability* that a photon will arrive at that point. In other words, while the photon is passing through the two holes in the screen they must think of it as a wave, but when it is arriving on the screen they must think of it as a particle.

To everyone's surprise it was proposed in 1923 that matter might behave in the same dualistic way. In that year Louis de Broglie put forward the idea that, just as light waves sometimes behave as particles of matter, so should particles of matter sometimes behave as waves. His suggestion was confirmed in 1927 when C.J.Davisson in America and G.P.Thomson in England showed experimentally that electrons, when reflected from crystals, behave as though they are waves. In passing, it is amusing to note that J.J.Thomson was awarded the Nobel prize for discovering in 1897 that the electron is a particle, and that his son, G.P.Thomson, was awarded the Nobel prize for discovering in 1927 that the electron is a wave!

We now know that all small particles of matter – electrons, protons, neutrons, and so on – behave as waves in certain experiments. For example, if the experiment in which light passes through two holes in an opaque screen is repeated using electrons, and the two holes are made very much smaller and closer together, the same curious effects can be seen. The electrons arrive on the second screen in a pattern similar to that observed with light; there are 'bright' bands where many electrons arrive, and 'dark' bands where there are few. Again it is impossible to understand or to visualize this behaviour in terms of classical science. To form these alternate bright and dark bands the electron must somehow or other be 'aware' of both holes in the screen and behave as an extended wave, and yet we know that in many other experiments the electron behaves as a very, very compact particle. Indeed it behaves as a 'point' with a radius of less than $10^{-15}$cm and, as such, cannot possibly be visualized as passing through the two holes at once. In order to interpret and predict this mysterious behaviour a new branch of mechanics has been developed – *wave-mechanics* – in which a particle is represented by a wave which tells us the probability of finding the particle.

We now accept the fact that both light (indeed all electromagnetic radiation) and matter behave either as particles or as waves depending on how we choose to observe them. Physicists can no longer offer us a satisfactory mental picture of what a beam of light, or an electron, is 'really like'; instead they give us a mathematical theory which predicts what a photon or particle will *do* or, to be more precise, what it will *probably* do, in any particular circumstance. It is, however, important to remember that although these two concepts – these metaphors of 'particle' and 'wave' – are apparently contradictory, they are still useful; they are in fact both rigorously valid within their own limited domains and allow us to predict precisely what will happen in the circumstances to which they apply.

Following a lecture by Niels Bohr in 1927, physicists call these two descriptions – wave and particle – *complementary*. In this lecture he said:

'The two views of the nature of light are to be considered as different attempts at an interpretation of experimental evidence in which the limitations of the classical concepts are expressed in complementary ways.'

This recognition that our picture of the physical world depends intimately on how we choose to observe it – and that, in that sense, the observer is part of the picture – marks one of the first points where modern physics exposed clearly the limitations of the Mechanical Philosophy and, in doing so, broke entirely new ground.

One of the principal ideas of classical physics was that underlying the subjective appearances of familiar things there is a real objective world which is independent of how we observe it, and that we can get to know about the properties of that real world through scientific observation. We now realize that science tells us, not about the intrinsic nature of things in that real, objective, world but it tells us about their extrinsic 'relations' to other things. In other words, modern physics does not tell us what atoms, electrons and photons are *really like in themselves*; it is content to tell us how they *behave* when we observe them. Apparently we must learn to think of a photon or an electron more as a particular form of behaviour than as a thing. To quote Bohr again:

'It is wrong to think that the task of physics is to find out how nature *is*. Physics concerns what we can *say* about nature.'

Let us now look at another strange idea which has come from the exploration of the very small.

## 2.15 **Chance replaces Certainty**

To make sense of the world we make constant use of the idea of cause and effect. Indeed, as that great mathematician the Marquis Pierre Simon de Laplace pointed out in 1814, it was a basic belief of the Mechanical Philosophy:

'We ought then,' he said, 'to regard the present state of the Universe as the effect of its antecedent state and as the cause of the state that is to follow. If for one instant a sufficiently vast intelligence were to know all the forces in Nature and the positions of all the bodies, then from one formula it could calculate the movements of everything in the Universe from the largest body to the lightest atom; both the future and the past would be present before its eyes.'

No doubt many people, including Laplace, knew that this could not be done in practice, nevertheless they did believe that it could be done in principle. Apart from adding electromagnetic forces to the list of things which the 'vast intelligence' would need to know, most 19th century scientists would have agreed with Laplace. In the 20th century most of us believe that he was wrong, not only in practice, but in principle as well; quantum theory has told us that we cannot predict the future of the 'lightest atom'.

To illustrate this let us return once more to the experiment in which a stream of electrons falls on two very small holes close together in a screen and passes on to a second screen. As we noted in §2.14 the electrons will arrive on the second screen in a pattern which shows that in passing through the holes they behave as waves. To calculate their behaviour, physicists use what they call 'wave-mechanics'; they associate a wave with each electron – the wavelength is proportional to the electron's momentum – and then treat the stream of electrons in much the same way as a beam of light. At any point on the screen the calculated intensity of this mysterious wave gives the probability, *not* the certainty, of finding an electron at that point.

There are many things about these 'waves of probability' which are very queer. In ordinary life, measures of probability represent the statistical behaviour of some quantity or event. They usually reflect our ignorance of the precise causes of an event, due perhaps to errors of measurement or to the sheer complexity of the causes. For example, we can consult the statistical tables prepared by Ladislau von Bortkiewikz to find the probability of being kicked to death by a horse in the Prussian Army between 1875 and 1894. But even if we had done this before joining up, it would have been impossible to assess whether or not we personally would survive. Nevertheless, we would still have believed, in company with the Marquis de Laplace, that causality applies even to the actions of a horse. Not so in wave mechanics! Quantum theory tells us categorically that our common-sense ideas about causality do not apply to the actions of an atom; in the atomic world different particles may do different things, although their initial states are identical. In other words, the quantum theory tells us that, in the microphysical world, *State A is not always followed by State B*, which contradicts flatly one of the basic assumptions of classical science. Indeed, the quantum theory has proved to be remarkably unlike the classical ideas of the Mechanical Philosophy in many ways. To derive the properties of an ensemble of particles in, say, a gas, classical mechanics starts with an individual collision between two particles, then deduces statistically what will happen in many such collisions, and hence calculates the properties of the gas. Quantum theory goes the opposite way; it starts by calculating the properties of an ensemble of collisions and then uses this result to predict what will *probably* happen in any one collision; furthermore, it tells us that we can *never* know what will *actually* happen.

When this remarkable idea was first put forward by the physicists who were developing the quantum theory in the late 1920's, it was not regarded as so startling as we are often led to believe. At that time, there was a widespread, post-war reaction against the strictly rational, deterministic aspects of science in the intellectual circles of German-speaking Europe where the quantum theory originated; the basic belief of the Mechanical Philosophy in the universal rule of cause and effect was being questioned even by some of the leading scientists. This general attitude to science can be seen in Oswald Spengler's *Decline of the West*, 1918, in which he inveighs against the tyranny of reason and points to the principle of causality as one of its major forms of expression.

Although the quantum theory was soon widely accepted by the scientific world, and has since proved to be an indispensable tool for analyzing atomic events, there are still basic questions about it which, for many people, have not yet been satisfactorily answered. Why, they have asked for 50 years, is the quantum theory limited to telling us what will *probably* happen; why can't it tell us what will *actually* happen? Most people, I guess, find this limitation easy enough to accept provided that they may attribute it to our ignorance; they prefer to think that, if we only knew more about the state of an atom, we should be able to predict with certainty what it would do. However, this is not what most physicists believe; on the contrary, most of them are convinced that no matter how hard we try, we shall never be able to predict individual atomic events! Indeed they regard our classical ideas about cause and effect as 'statistical' approximations which are based solely on our experience of the behaviour of the objects which we meet in everyday life, objects which are large compared with atoms. In their view we can predict what will certainly happen when one billiard ball collides with another only because the billiard balls are made up of such a vast number of atoms. The statistical uncertainty in the behaviour of a billiard ball is so small that it approximates to certainty, and only in that sense can classical physics tell us what will certainly happen.

As one might expect there have been many worthy opponents of this point of view. The best known was Albert Einstein who refused to believe that 'God plays dice with the world'. Writing to Niels Bohr in 1924 he said:

'I cannot bear the thought that an electron exposed to a ray should by its own free decision – *aus freien entschluss* – choose the moment and the direction in which it wants to jump away. If so, I'd rather be a cobbler or even an employee in a gambling house than a physicist.'

Broadly speaking, Einstein argued that the failure of quantum theory to predict what would certainly happen in an individual atomic event did not mean that these events were governed by pure chance; to him it meant that the quantum theory itself was incomplete.

Ever since it was put forward the quantum theory has been defended vigorously against all attempts to preserve the classical ideas of cause and effect. One line of defence has been to argue that any attempt to do so is meaningless. Causality, it is argued, necessarily implies that to predict the future with certainty we must be certain about the present, and for quantum events that is prevented by Heisenberg's *Uncertainty Principle*.

In 1927 Werner Heisenberg pointed out that uncertainty is an unavoidable feature of all our knowledge of physical events. Suppose, for example, that we wish to know the position and momentum (momentum = mass × velocity) of a particle such as an electron. In principle we can measure its position as accurately as we like by illuminating it with radiation of sufficiently short wavelength; the shorter the wavelength the more precisely we can 'see' where it is. However in illuminating the electron we necessarily disturb its path and thereby make subsequent measurements of its momentum correspondingly uncertain; furthermore, the shorter the wavelength we use, the more we disturb the electron. Thus, as we decrease our uncertainty about its position, we necessarily increase our uncertainty about its momentum. It is simple to show that the product of these two uncertainties can never be less than Planck's quantum of action.

It follows from this argument that, no matter how hard we try, we cannot hope to measure what we need to know about an individual atom with sufficient precision to predict what it will *certainly* do. In practice, individual atomic events are therefore bound to remain largely unpredictable, and so to discuss whether or not they are unpredictable in principle is a waste of time. Heisenberg called it 'sterile and senseless'.

A second line of defence has been the effort to prove theoretically, that the unpredictability of atomic events is not due simply to our ignorance of all the factors which govern them, but is an inescapable feature of nature. The first of these proofs was put forward by John von Neumann in 1932 and since then there have been others. I doubt whether any of them can be regarded as conclusive. Reading them, I am strongly reminded of the endless arguments in the 19th century about the reality of a luminiferous aether in which all attempts to picture the world in familiar mechanical terms were eventually defeated.

As a last example of the queer ideas which have been encountered in the exploration of the very small, we shall now look at an experiment in physics which illustrates dramatically the limitations of our classical, common-sense ideas about the world.

## 2.16 **A Failure of Reductionism – some things cannot be taken to pieces**

'Without dissecting and anatomizing the world most diligently', wrote Francis Bacon in 1620, 'we cannot found a real model of the world in the understanding'. For more than 300 years scientists have followed Bacon's advice and it has paid rich dividends. Indeed, apart from the basic faith that the world can be understood in terms of cause and effect, the idea that the best way to understand things is to take them to pieces is the most characteristic feature of the classical 'scientific' approach. Needless to say this so-called 'reductionist' approach of science has always had its critics, and its limitations have been debated for centuries.

In recent years an interesting new light has been thrown on this old debate by quantum physics which has shown us that, even in the 'mechanistic' world of physics, reductionism has its limits. The story starts with a paper written by Einstein, Podolsky and Rosen (EPR) in 1935 [1]. In this paper they drew attention to the surprising fact that, in certain experiments in which two particles or photons have a common origin, quantum theory predicts that the result of a measurement on one of these particles or photons at a place A will depend on the result of a measurement on the other at a place B, *although A and B may be remote from each other and there appears to be no way in which the two particles or photons can 'communicate' with each other.* Einstein referred to this effect as 'spooky action at a distance', and it is commonly called the *EPR paradox.*

The apparatus sketched below illustrates this paradox. A source of light (S) emits pairs of photons which travel in opposite directions. Because each pair of photons is emitted by a single atom (e.g. an excited mercury atom) in the source, both the photons of a pair always have identical polarization and the plane of this polarization is randomly distributed. In the path of these photons we put polarizers (P1, P2) which accept light which is polarized in one plane and reject light which is polarized at right angles. The two polarizers are set

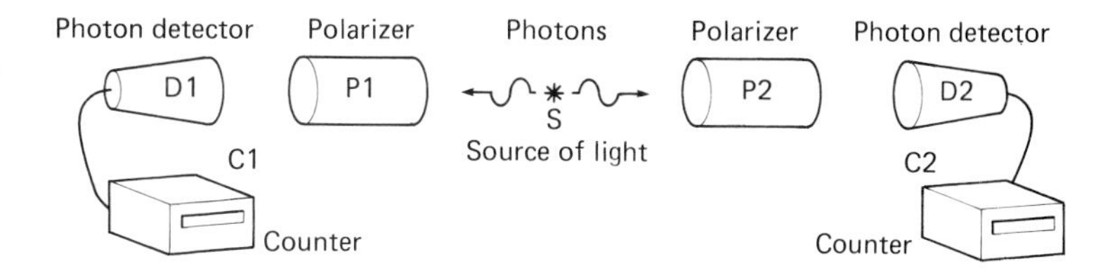


A practical version of the Einstein–Podolsky–Rosen paradox (EPR).

at right angles to each other. Beyond each polarizer there is a photon detector (D1,D2) and a counter (C1,C2) which registers a count every time a photon arrives.

Let us now consider what happens in two separate cases; firstly when the polarization of the photon pair is parallel to one of the polarizers. In this case, common-sense, classical science and quantum theory all tell us that the photon which is polarized parallel to the polarizer, say P1, will pass through and be counted, while the other photon will find itself at right angles to the polarizer (P2) and will be stopped. Thus a photon will be counted by C1 but not by C2 and so there will be no coincident counts.

Now consider what happens when the photon pair are polarized at exactly 45 degrees to the two polarizers. The classical 'reductionist' approach tells us that each photon will have an even chance of passing through its respective polarizer, and so, if the source emits N pairs of photons per second and the detectors are 100 percent efficient, then N/2 photons per second will be counted in each counter. Again, taking a common-sense view in which each photon of a pair has an equal and independent chance of passing through its respective polarizer, we would expect half of the pairs to be counted and the two counters (C1,C2) to count N/4 coincident counts per second. But quantum theory tells us that we are completely wrong! It tells us that if one photon gets through its polarizer and is counted, its twin will never go through the other polarizer and be counted, even though it apparently has an even chance of doing so. Thus quantum theory predicts that, in this case, there will be no coincident counts at all.

It seems that, according to the quantum theory, one photon of a pair must always 'make up its mind' to do the opposite of its twin; furthermore – and this is even more unexpected – the theory tells us that the system will behave in this same queer way even if the distances of the two detectors from the source are made unequal, or if they are made so great that we can no longer imagine the two photons to be interacting with each other. In any attempt to explain these predictions of quantum theory in common-sense terms we are faced with the difficulty that the photons of a pair behave as though they were in touch with each other, and yet there is no apparent way in which this communication can take place!

However, according to the established interpretation of the

quantum theory there is nothing to worry about and there is no paradox. The two separated particles or photons have a common origin and must be treated, not as self-existent, separate, entities with independent properties, but as a *pair*. In the jargon of quantum mechanics, their interaction with the apparatus is described by a 'joint wave-function', and we must think of the two photons as forming one single entity together with the whole apparatus which detects them – *we cannot 'reduce' or explain the way the system behaves in terms of the independent properties of its constituent parts.*

It is not surprising that this so-called paradox has worried many people ever since it was first published in 1935. Einstein claimed that it showed that there must be other factors which we don't yet know, 'hidden variables', which govern the behaviour of particles and photons, and so the the quantum theory is not wrong, but must be incomplete. Since then, many unconvincing attempts have been made to explain the EPR paradox in terms which can be understood 'mechanically'. It has been suggested, for instance, that the two photons may interact, perhaps instantaneously, through the medium of some unknown universal field; again, it has been suggested that we may have to revise our ideas of 'locality' so that we no longer think of objects as existing only in separated spaces, and so on ....

In 1964, largely due to the work of J.S.Bell [2], it was realized that the EPR paradox offered an opportunity to compare, experimentally, the predictions of quantum theory with those of the attempts which have been made to solve the paradox in classical terms, including those which invoke 'hidden variables'. Since then, several elaborate experiments have been made, both with particles and photons, and they have shown, beyond reasonable doubt, that it is the quantum theory, and not classical physics, which gives the right answer.

Evidently we must accept that this strange behaviour of photons (or particles) is just one more example of the behaviour of things in the physical world of which we cannot form a satisfactory mental picture or 'mechanical' explanation. In their protests against this attitude many people are apt to forget that many of the conventional theories offered to us by physics, such as those which we accept as explanations of the nature of light or the force of gravity, were once rejected by many scientists as being 'occult' – or in Einstein's words as 'spooky' – because they could not be understood in common-sense terms. What they fail to realize is that our discovery of the nature of the physical world owes more to experience than to theory.

The moral of this story is that 'reductionism' has limits. One of the central assumptions of the Mechanical Philosophy was that nature could be understood by taking it to pieces, and that the operation of any system could be understood in terms of component parts with intrinsic properties which were independent of the rest of the system. What we have now found out is that this idea has limits. We have discovered that there are some things in nature which cannot be understood in this way, they must be treated as a whole.

## 2.17 **Exploring the very large – a new understanding of Space and Time**

Two notable events in the first decade of the present century warned scientists that the physical world is not as straightforward as they had previously thought. The first was when Max Planck put forward the idea of a quantum of energy (§2.13) at a meeting of the German Physical Society in 1900; the second was when Albert Einstein published his first paper on the *Special Theory of Relativity* in 1905.

In his *System of the World* Newton had set the solar system in a solid framework of absolute time and absolute space:

> 'Absolute space', he wrote, 'exists without relation to anything external, it remains always similar and immovable; absolute time flows equably without relation to anything external.'

Nevertheless, it was impossible, he pointed out, to measure our movement with relation to this framework of absolute space by observing 'the position of bodies in our regions', because we have no way of knowing which of these bodies is absolutely at rest. He went on to argue that the laws of motion would be the same for all observers in uniform motion.

But to the physicists of the 19th century it seemed that, although the laws of motion might well be the same for all observers in uniform motion, the laws of electromagnetism were not. If light is propagated as a wave in an aether which pervades all space, then it should be possible to detect the motion of the Earth through this aether by measuring the apparent velocity of light in different directions on the Earth. When light is travelling through the aether in the same direction as the Earth, then its velocity, as measured on Earth, should be less than when it is travelling the opposite way; just as the speed of a boat as seen from the bank of a river appears to be slower upstream and faster downstream.

The trouble with this beautifully simple idea was that, when it was put to the test, it was found to be wrong. In 1887 Michelson and Morley showed experimentally that the velocity of light is the same in *all* directions on the surface of the Earth; they found no sign of an aether drift. This unexpected result was, to quote Lord Kelvin, one of the 'two clouds over physics' at the end of the 19th century, and it did not disperse until Einstein published his paper on the *Special Theory of Relativity* in 1905. In that paper Einstein threw Newton's solid framework of absolute time and space out of the window and put forward a theory in which all the laws of physics – not only the laws of motion – are the same for all observers no matter how fast or in what direction they may be moving, provided only that their motion is uniform.

A very strange assumption which is central to this theory is that all observers in uniform motion measure the same speed of light, irrespective of whether they are moving with respect to the source of light or the source is moving with respect to them. This simple but revolutionary idea was certainly consistent with the failure of Michelson and Morley to detect the Earth's motion through the aether, but it was, of course, completely contrary to common-sense. It implied that the whole analogy between light waves in the aether and sound waves in the air, or waves on water, must be wrong. In Einstein's new picture the aether had no 'mechanical' properties whatever, and so physicists had to accept the unwelcome idea that if they wished to think of light as a wave, then they had to picture this wave as waving in a medium which was indistinguishable from empty space. In other words they had to accept that a satisfactory mental picture of light cannot be made by a simple analogy with the behaviour of familiar things.

There are many other strange effects predicted by the Special Theory of Relativity which illustrate the central idea that the facts of physics are observed facts and, as such, depend on exactly how they are observed. For example, a simple analysis shows that the measured mass of an object and the rate of a clock are all relative measurements, and depend on how fast the object and the clock are moving with respect to the observer. Thus, as the object moves faster, its mass increases and its length in the direction of motion decreases, while the rate of the clock decreases. The same analysis shows that the increase in mass of a moving body is exactly proportional to the increase in its kinetic energy, and that we cannot distinguish between a gain in mass and a gain in energy. This result was generalised by Einstein as the equivalence of mass and energy in a formula which everybody knows $E = mc^2$. These relativistic effects are only significant at velocities which are an appreciable fraction of the speed of light and are therefore completely negligible at the speeds which we commonly encounter in everyday life. On the other hand they are most important at the very high speeds which are commonly encountered in the study of atomic particles in accelerators and cosmic rays, in nuclear engineering and in all physical processes where very high velocities are involved. At first sight it is tempting to regard them as nothing more than illusions, rather like the familiar illusion that people appear to grow smaller as they walk away. We prefer to think that the mass of an object and the rate of a clock cannot 'really' change with speed, and that they only 'seem' to do so. An instructive illustration of the pitfalls in thinking in this way is given by the 'twin paradox' which, although hackneyed, is highly instructive.

Twin brothers live on Earth and both possess identical clocks and calendars. One fine day one of the brothers sets off in a space ship which travels at about 260,000 km per second to take a look at a nearby star, taking his clock and calendar with him. The travelling twin takes careful account of the passage of time and marks off a new day on his calendar every 24 hours as measured by his clock. After 10 years by his calendar, he returns to Earth and finds that, according to his brother's calendar, he has been away 20 years. This cannot be explained by saying that time on the spaceship only *appeared* to go slower than time on Earth; the travelling twin is now *actually* ten years younger than his brother, even his hair has not gone grey. What he has done, according to the theory, is to take a different route through time as well as space.

This story is called a paradox because, at first sight, there appears to be no good reason why one twin should get older than the other; it seems as though we ought to be able to look at the problem from the point of view of whichever twin we choose and regard the motion of the other one as relative. An explanation of what is wrong with that point of view is to be found in most text-books on relativity and is too long to be given here; briefly it boils down to the argument that the situation is not truly symmetrical; one twin is in accelerated motion and the other is not. I should add that although the experiment has never been done with twins, there is plenty of experimental evidence to show that the theory is correct; it has been shown, for example, that the lifetime of atomic particles travelling with high velocity is increased by the amount which the theory predicts, and experiments have also been done with very precise clocks carried in aircraft.

The twin paradox illustrates dramatically that our conventional picture of time (Newton's time) as something which flows past us at an unchanging speed, and of space and time as completely independent, must be wrong. By taking a different route through space the travelling twin has taken a different route through time; evidently time and space are not as independent as we had thought. The idea that space and time are not independent dimensions, but form a four-dimensional continuum which we now call space-time, was first put forward by Hermann Minkowski in 1908.

A major contribution to our understanding of the structure of this space-time was made by Einstein when he published his *The General Theory of Relativity* in 1916. In this remarkable work Einstein removed the last traces, at least in physics, of the ancient idea that events on Earth are in some way different to events in the sky. He extended the *Special Theory of Relativity* to tell us that the laws of physics are the same for all observers, no matter where they are or how they are moving. His new theory was no longer restricted to uniform motion in a straight line, but included the effects of accelerated motion and gravitation.

The *General Theory of Relativity* has proved to be essential to our understanding of things which are very large. It has told us a lot about the structure of the space-time in which we live; it has shown us, for example, how space, time, mass, inertia, gravitation, the paths of light rays and many other things, are related in this space-time. It is, in fact, vital to our understanding of all celestial phenomena which involve very high energies, intense gravitational fields, and very great distances. It allows us, for instance, to calculate what goes on in very dense stars, such as white dwarfs, neutron stars, black holes and so on. Again, it is essential to all our attempts to make models of the structure of the Universe and to relate what we see in our large telescopes to those models.

The *General Theory of Relativity* has turned cosmology into a branch of very difficult mathematics and geometry; Pythagoras and Kepler would have liked it. It is a pity that the theory is so mathematical that most people, including many scientists, cannot fully appreciate what an important step forward it represents in our understanding of the physical world and of the structure of the Universe. It is one of the major intellectual achievements of modern science.

### 2.18 **Exploring the very large – the realm of the Galaxies**

There are three major differences between our modern ideas about the Universe and Newton's *System of the World*; we believe that it is much bigger and much older, and we believe that it evolves with time. As we have enlarged our telescopes so we have enlarged our picture of the Universe. When telescopes were only a few centimetres in diameter (as they were in the 17th century), we recognized that the Earth is a planet of the Sun and that the stars are many and far away. By the end of the 18th century, when telescopes were built with a diameter of about 50cm, we realized that the Sun is only a minor star in a Galaxy of stars; later we discovered that this Galaxy has a population of about 100 billion stars and a diameter of 100,000 light years.

When, early in the 20th century, the diameter of the largest telescopes reached 2.5m we learned that our Galaxy is itself a member of a Universe of galaxies extending as far as we could then see, roughly 500 million light years. Furthermore, we learned the surprising fact that the Universe is expanding and that the other galaxies are receding from us with a velocity which increases with their distance. In the last 50 years, telescopes have doubled in size, and it won't be long before they double again; even larger radio-telescopes have been built and we have seen further into space. We can now see galaxies whose light has taken roughly billions of years to reach us and we realize that there must be at least 10 billion of them.

Our current interpretation of this new picture, with its vast stretches of space and time, tells us that the Universe was created between 10 and 20 billion years in the past, that the Sun and planets were formed about 4.5 billion years ago and, according to anthropologists, that man-like creatures have only been on Earth for about 3 or 4 million years. Man has only been around for less than one thousandth of the age of the Earth! Modern science has certainly made the Earth look very much smaller and human history very much briefer than they looked in medieval times, or for that matter, in Newton's *System of the World*.

One obvious consequence of this enormous increase in the size and age of the visible Universe has been a revival of interest in the old idea that there may be life on other worlds. Ever since it was accepted that the Earth is a planet and that the distant stars are bodies like our Sun, it has been argued that among all the stars there must be some with inhabited

A group of galaxies (Stephan's quintet).

planets. In recent years this argument has become more persuasive as our estimates of the size and age of the visible Universe have increased. Many plausible attempts have been made to estimate how many stars might have habitable planets, but the weak link in the chain is the difficulty of estimating how many of these habitable planets are likely to be inhabited. Although there are plenty of theories, as we shall note later, we don't really know how life was formed on our own Earth and, until we do, we can't estimate the possibility that it has been formed elsewhere.

We can, of course, guess the probability that life will arise, or has arisen, on a planet, and most attempts to do this suggest that it is a wildly improbable event; we can then multiply our guess by our guess at the number of habitable planets. Many people have done this and written long papers about it; their estimates of the number of inhabited 'worlds' in our Galaxy vary from 0 to at least 10,000! Nevertheless, the general argument that it is worth while to keep a lookout for other civilisations is persuasive, and in recent years the search for extra-terrestrial life by listening for radio signals has become scientifically respectable. Indeed in 1982, the International Astronomical Union established a new Commission (No. 51) on the Search for Extra-terrestrial Life.

## 2.19 **In the Beginning**

The history of Newton's *System of the World* was simple; it was created by God in its present form at some time about 4000 BC and would end when He so decided. In due course the Biblical time scale was thrown out when the study of fossils, rocks and biology in the 18th and 19th centuries showed that the Earth itself and the creatures that live on it have evolved over an immensely long history which goes back for millions of years. In the 20th century an attempt has been made to re-write, so to speak, the *Book of Genesis* to suit this new time scale. The story is by no means complete but there is enough of it to be interesting. Let us start at the beginning and look at what modern science has to tell us about how the Universe began. In the last 30 years or so, there have been two competing cosmologies – the *Evolutionary* and the *Steady-State*, which differ fundamentally in the way their models of the Universe change with time. In the evolutionary cosmology, the large-scale properties of the Universe (its radius, density, etc.) change with time, and it makes sense to talk about a beginning and an end. In the steady-state cosmology, the visible Universe always looks much the same and so there is no beginning and no end. Both these cosmologies must necessarily account for the fact that the Universe is expanding.

In the most popular version of the evolutionary cosmology, the Universe is pictured as expanding away from an initially compact state – the 'primeval fireball' – in which the temperature and density were almost infinitely high. This alarming picture of Creation has come to be known as the Big Bang, and it is estimated that it took place between 10 and 20 billion years ago.

In the steady-state cosmology the Universe expands but, to keep its density and local properties unchanging, new matter is constantly created everywhere; on an average, the mass of one hydrogen atom appears spontaneously in each litre of volume every 500 billion years. Thus by the continuous creation of matter, the steady state cosmology avoids the problems of how the whole Universe started and of how it will end. This rate of creation is, by-the-way, far too low to be detected in the laboratory.

Since the steady-state cosmology was put forward in 1948 a major effort has been made to test these rival cosmologies by comparing them with observations and, at the present date, most cosmologists favour the Big Bang. There are at least three reasons for their preference, which for a cosmological debate represents strong evidence, cosmology being a branch of science in which too many theories chase too few facts!

Firstly radio-astronomers have found that the farther out they look into space, and hence the farther back in time, the greater appears to be the density of radio-sources. In other words, more radio-sources must have been formed in the past than are being formed now. This conclusion points to an *evolving* Universe, not to one in a steady state. Secondly it was observed in 1965 that there is a faint background of radiation at very short radio-waves, microwaves, received on Earth from all directions in the sky. This discovery was immediately welcomed as evidence in favour of the Big Bang because it had been predicted, some 30 years before, that the intense radiation generated by the Big Bang whould still be observable as a weak background of radiation from the sky. One can't help wondering, by-the-way, whether this background radiation might not have some other, perhaps less exotic, explanation, but so far none has been found. The third piece of evidence is that the ratio of hydrogen to helium found in our Galaxy is in reasonable agreement with the ratio to be expected theoretically if they were both formed in the Big Bang.

These three facts fit well with the generally accepted theory that 'in the beginning' the Universe was immensely compact, and probably immensely hot, and the problem now is to make this theory fit all the other facts. One fact which has to be explained is that, on a very large scale, the Universe looks to us to be remarkably uniform in all directions. It is, for instance, difficult to explain how the Big Bang could have produced a background of microwave radiation with an intensity which is the same, within at least one part in ten thousand, all over the sky. Any explanation of this uniformity has to be reconciled with the highly non-uniform way in which, on a smaller scale, the population of galaxies is distributed in clusters and sheets.

Another interesting question is why the Universe should be expanding at the actual rate which we observe some 10 or 20 billion years after the Big Bang? The problem is to show how this rate can be reconciled with a model of the Big Bang which does not require the initial rate of expansion to have been fantastically precise. If the rate had been too fast the Universe would have dispersed before the galaxies could have been formed; if the rate had been too slow the Universe would have collapsed before now. At the present time one suggestion, which is attracting a lot of attention, is the so-called 'inflationary' Universe. It solves many of the problems encountered by the older theory of the Big Bang by invoking some of the latest ideas about the properties of matter at extremely high temperatures. This brings us to the reason why, nowadays, physicists are so interested in the theory of the Big Bang.

One of the central aims of physics is to understand the relationship between the four principal forces which we find in nature – gravity, electromagnetism, and the 'strong' and 'weak' nuclear forces which act between atomic nuclei. During the 1960's the first step towards unifying our ideas about these forces was taken when it was predicted that the electromagnetic and 'weak' nuclear forces are related, and that they might prove to be indistinguishable in nuclear reactions at energies exceeding about 100 Gev. This prediction has recently been supported by experiments with very energetic particles produced in a colossal accelerator at CERN in Geneva. It has now been proposed that the 'strong' nuclear force might prove to be indistinguishable from both electromagnetism and the 'weak' nuclear force in events with immensely higher energies ($10^{14}$ Gev), and that all four forces, including gravity, might be indistinguishable from each other at even higher energies ($10^{19}$ Gev). There is obviously no hope of verifying these predictions in the laboratory, because they demand energies well beyond the capabilities of any conceivable accelerator; the only place where such colossal energies might have occurred in nature is in the Big Bang when the Universe was only $10^{-35}$ seconds old, and that is one reason why physicists are now taking a lively interest in cosmology, and cosmologists are taking a lively interest in high-energy physics. By studying the early stages of the Universe, the physicists hope to throw light on the behaviour of matter at immensely high energies and to test their grand unifying theories (inevitably called GUTs) by which they seek to unify their ideas about the forces of nature; by studying those new theories the cosmologists hope to gain a better understanding of the Big Bang.

If we now ask cosmologists how the primeval fireball came into existence, or in other words, what happened before the Bang, they tell us with commendable modesty that they don't know. Our present knowledge of the laws of physics fails at the stupendous densities• involved and, not surprisingly, cannot be used to describe what happens in the very early stages – before $10^{-43}$ seconds! – of the Big Bang. Furthermore the question is mildly embarrassing to many of them because it seems to involve creating something out of nothing (*creatio ex nihilo*), and looks like one of those questions about first and last things which they regard as being outside the boundary fence of science – something perhaps for the theologians. We seem to have reached a point where the modern *Book of Genesis* has nothing to add to '*Fiat Lux*'.

Maybe the cosmologists have lost their ball to the theologians for the time being, but some of them are starting to look over the fence to see if they can get it back. One idea which they have come up with is that the 'creation' of the Universe may be analogous to the 'creation' of the so-called 'virtual' particles of modern physics. According to quantum theory, 'virtual' particles appear spontaneously in space with a lifetime which depends on their total energy. An electron-positron pair, for instance, can appear out of nothing and last for $10^{-21}$ seconds. A system with zero total energy could, or so it is argued, appear spontaneously out of nothing and last for ever. If the total energy of our Universe can be regarded as zero – the total positive energy in the mass may be equal to the total negative potential energy in gravitation – it might, like a 'virtual' particle, have appeared spontaneously out of nothing! To most people, I guess, that would seem to be more like a conjuring trick than an explanation. However it may be that all our efforts to 'explain' the world end up, like the 'bootstrap' theory (§3.9) of fundamental particles, in circles in which one mystery is explained in terms of another. Maybe that is inevitable because we ourselves are part of the system which we are trying to explain.

So much for the scientific version of the first two days of Creation – but what about the rest of the week? For the third and fourth days the story is reasonably coherent, although rather patchy. Astrophysicists can tell us quite a lot about how the stars were born out of the interstellar gas in galaxies and how these stars might have formed planets like the Earth, but they can't tell us how or when the galaxies themselves were formed out of the expanding Universe. There are many ideas about how this happened, but no one has yet seen a galaxy actually in the process of formation. Perhaps we shall not have long to wait; a powerful new telescope is shortly to be launched into space where it will get a clearer view of the sky than any other telescope has ever had before, and there are other very large ground-based telescopes under construction.

It is when we come to the fifth and sixth days of Creation, when the *Book of Genesis* tells us that God created living creatures and man in His own image, that the story has several missing links. The first missing link – the one that gets the most attention – is the formation of the self-replicating protein molecules, the basic components of life. How and where these molecules were formed and, in particular, how long such a process might have taken, is not known. Some people will tell you that it is quite plausible that they were formed in the primeval slime from the basic elements of carbon, hydrogen, oxygen and nitrogen. Some will tell you that such an event is wildly improbable, comparable to the formation of a Boeing 747 by a high wind blowing through a junk yard. If they were formed by a random process, so it is argued, then they must have been formed outside the Earth where there would have been more space and time. This is a lively and, to my mind, an entertaining argument which as far as I can see is likely to go on for a very long time, at least until we discover some facts about life in space.

There are, however, a number of other less obvious links missing from the scientific version of *Genesis*. For instance it has been pointed out that the Universe has many properties which are essential to life and yet which cannot be explained as the necessary consequences of the known properties of matter; nor it seems, can they be comfortably attributed to pure chance. Let us look briefly at some of these 'happy accidents'. First of all there are all the obvious considerations. Life as we know it requires a special environment, a certain range of temperature, an atmosphere, the presence of elements such as carbon, and so on. In turn this environment demands a star, our Sun, which is neither too hot nor too cold and a planet which is large enough to retain an atmosphere by the force of its gravity, and yet not so large that it becomes a star in its own right. These requirements are all well known and most scientists would agree that they can be explained in terms of chance. There are so many billions of stars and galaxies and so much time for things to happen that, even if the formation of living things is improbable, it would have happened somewhere by chance.

However, this argument from chance is not quite so convincing when we look at some of the other 'happy accidents'. It has been pointed out, for instance, that a change of a few per cent in the strength of some basic forces, such as gravitation or the strong nuclear force, would have made the formation of life impossible. If the strong nuclear force were slightly weaker than it is, then the deuteron (the nucleus of heavy hydrogen) would not exist, and the deuteron is essential to the nuclear reactions which supply energy in stars; on the other hand, if the strong nuclear force was slightly stronger, then two protons could join to form a diproton which would make the existing nuclear reactions in stars unstable. As another example, astrophysicists point to the formation of carbon which is essential to the sort of life we know. The carbon nucleus is formed in elderly stars by the fusion of three helium nuclei. For this to happen there has to be a favourably placed resonance in the carbon nucleus, and to prevent the destruction of the carbon there has to be another favourably placed resonance in the oxygen nucleus. If these two resonances had slightly different values carbon could not have been formed.

Cosmologists also have their 'happy accidents'. They point, for instance, to the curious 'accident' that the life-time of the Sun ($10^{10}$ years) is comparable with the time-scale of biological evolution, and yet these two time-scales do not appear to be related through the properties of matter.

At the present time these 'happy accidents', which have conspired to produce the world we live in, seem to some people to be too elaborate, too unlikely, to be satisfactorily explained by pure chance. It may well be that as science advances, they will be explained as being a necessary consequence of the basic properties of matter; in the meantime, the only explanatory idea which has been offered is the so-called 'anthropic principle'. Quite simply this so-called 'principle' draws attention to the idea that, although we no longer see life as being central to the Universe, as it was in the Medieval Model (§1.2), the opposite idea, that life is in no way relevant to the Universe, may be equally untrue. It offers an 'explanation' of those highly improbable features of our environment which are favourable to life by pointing out that, had they been otherwise, we shouldn't be here to observe them. In that sense, the 'existence' of our Universe can be said to depend upon it producing an observer! We are here because we are here.

There is, I should mention, another way of facing the improbable features of our Universe and that is to postulate that it is only one of many possible worlds. According to this idea, these other worlds exist and embody all the possible ways in which the Universe might have been made; our particular Universe is then no longer improbable, it just happens to be the one in which we happen to be. This idea was suggested by the role of probability in quantum theory. I find it hard to take it seriously, and once more I am reminded of those medieval arguments about angels dancing on the head of a pin.

I am also reminded of the old Argument from Design which in the 16th and 17th centuries was one of the major proofs of the existence of God. It seems to me that the 'anthropic principle' may be this old argument in a new, more esoteric, disguise. To quote one eminent scientist:

'A common sense interpretation of the facts suggests that a super-intellect had monkeyed with physics as well as chemistry and biology, and there are no blind forces worth speaking about in Nature.'

That was written a few years ago; it could just as well have been written in the 17th century.

The *Book of Ecclesiastes* tells us:

'He hath set the world in their hearts, so that no man can find out the work that God maketh from the beginning to the end.'

Perhaps that will eventually prove to be so, but in the meantime it is one of the major challenges of the adventure which is called Science to see how much we can find out about the beginning.

In the first chapter of this book we saw how science has been used, in the words of Marx, to 'change' the world and, having been so successful, has come to be largely identified with its applications – it has become a modern cargo cult. Its practical success has confirmed Francis Bacon's forecast that 'knowledge and power meet in one', and has shown beyond doubt that this power can be applied 'to endow the condition and life of man with new power and works'. It has established a widespread belief that progress is more or less inevitable in a 'scientific society' and can be assured by the practical application of knowledge.

In this second chapter we have seen how science has been used to 'interpret' the world. We have seen how it destroyed the early cosmology in which the Earth was at the centre of the Universe and in which there was a fundamental difference between things celestial and terrestrial. From the time of Descartes (1596-1650) until the end of the 19th century it offered a confident interpretation of the physical world in the common-sense terms of the Mechanical Philosophy. In the present century, however, we have begun to understand the limitations of that philosophy. We have found that common-sense cannot be used to understand the atomic and sub-atomic world, and that our everyday experience of things that we can touch and see is only a small part of the story. As one example, we now recognize that we cannot know what things are 'like in themselves', but only how they 'behave' when they are observed; as another we have had to exchange certainty for probability in predicting what will happen in atomic events. At the other end of the scale, astronomers have shown us that the Universe is much vaster, more abundant and more wonderful than anyone could ever have imagined.

Both physics and astronomy have shown us that the Universe is a far stranger place than was ever envisaged by the mechanical philosophers.

In the next chapter we shall look at how science can, and should, be used to enrich our society.

## References

1. 'Can quantum-mechanical description of physical reality be considered complete?', by A.Einstein, B.Podolsky and N.Rosen, *Physical Review*, **47**, 777-780, 1935.
2. 'On the Einstein-Podolsky-Rosen paradox', by J.S.Bell, *Physics*, **1**, 195-200, 1964: *Review of Modern Physics*, **38**, 447, 1966.

# 3 The Cultural Dimension of Science

'The body of technical science burdens us because we are trying to use the body without the spirit.'
Jacob Bronowski

## 3.1 Our Image of Science

If we were to ask those legendary oracles of our society, the man and woman in the street, whether science is really worth bothering about, what would they say? Three hundred years ago they would, most likely, have never heard of it, and if they had, they would have regarded it as a hobby of the idle rich. One hundred years ago, they might have said: 'Yes, of course it is! It has made our lives more comfortable and interesting; it has given us better health, faster and easier travel, improved communications, and has brought to us a variety of goods, services and entertainment such as ordinary people have never seen before in history.' Almost certainly they would have taken it for granted that the improvement of our material standard of living is a worthy social objective and they would have looked to science to make it happen.

As we saw in §1.2, this idea of progress has only become part of our conventional wisdom in the last few hundred years. It dates from the time when European thought emerged from the Middle Ages, shook off a belief in magic and the intellectual authority of the Church, reasserted the power and potentialities of human reason and initiative, and turned its face from the past to the future. To this historic process we attach convenient labels such as the Scientific Revolution, the Enlightenment, the Industrial Revolution. The idea that 'knowledge is power', power to make material progress, was spelled out by Francis Bacon early in the 17th century (§1.2) and the subsequent history of science has proved him to be right. Indeed so successful have the applications of science been that the effort devoted to scientific research and development in the 'developed countries' has grown from almost nothing a century ago to about 2 per cent of their gross national product at the present time. In the process we have accepted that applied science is the mainspring of our material progress and, following Bacon, we have come to equate the value of science with its practical benefits.

To-day the answers to our question about the value of science are less optimistic. Most people, as surveys of public attitudes to science in the USA show, still hold science in high regard and believe it to be essential to progress. However, although they value it for its practical benefits – especially medical science – they regard it as a mixed blessing. They are aware that, although our science-based industries have brought them a wide variety of goods and services, some of

One hundred years ago the scientist was seen as unquestionably benevolent (an advertisement guaranteeing the purity of Cadbury's cocoa).

these industries have polluted our rivers and seas and are consuming natural resources which cannot be renewed; that automation has relieved drudgery, but at the cost of unemployment; that improvements in medical science have reduced deaths from disease which, in many parts of the world, has lead to over-population and so malnutrition; that the mechanisation of agriculture has produced more food but has caused a drift of populations away from the countryside into overcrowded towns. In the 'developing countries' people who take an interest in politics might add that science is not simply an agent of progress, but is an instrument which serves the political and economic interests of the 'developed countries' and their transnational corporations.

In most people's minds the worst effect of applied science is the enormous increase in the elaboration and destructive power of modern armaments. We have all heard the insane statistics. The two major powers have the capacity to destroy each other many times over; in the mid-1980's the expenditure of the world on armaments is more than three billion dollars a day, roughly 6 per cent of the total domestic product, while a large fraction of its population is undernourished; the 'developing' countries spend more than three times as much on armaments as they do on health care; one half of all the research and development in the world is devoted to armaments; three quarters of all spacecraft have been launched for military purposes and so on *ad nauseam*. No wonder the man and woman in the street take a rather dim view of science.

As well as criticizing the practical misuses of science, a few of our men and women in the street would complain about its cultural influence. (Although I distrust the word culture, because it has overtones of folk dancing, chamber music and the refinement of taste, I shall use it here for want of a better word, but only in the widest possible sense to embrace all our beliefs about the world and how best to live in it.) Their complaints will probably have been inspired by the well-publicized attacks on science by the so-called 'counter-culture'. Like Old Testament prophets coming out of the wilderness to warn the world against sin, the spokesmen of the counter-culture have been coming out of the wilderness of industrial civilisation in recent years to warn us against science. Armed with Wordsworth and William Blake, they tell us that science is a spiritual cul-de-sac and that if we want to save our souls and our society we must back away from it.

Science, so its critics say, has taught us to see the world as an impersonal machine which can best be understood in terms of its component parts and of measurable quantities like mass and velocity. So successful has this analytical view proved to be, especially in advancing the physical sciences that it has focussed the attention of our culture on a limited, 'mechanical', part of our experience at the expense of other more vital, more 'human', parts which have been left in the shade. This 'single vision', so they say, has been damaging to society because many of our most pressing problems, particularly social problems, cannot be solved by the analytical methods of science and must be seen as a whole. Furthermore, by concentrating our attention on quantities rather than qualities science has narrowed our imaginations, and by its concern with things and not people it has encouraged us to treat people as things, and thereby has impoverished our whole attitude to human relations as well as to Nature. Many of these critics add that, by undermining belief in religion,

The wildernesss of 19th-century industrial civilization. Clarke's Anchor Thread Works at Paisley.

science has desacralized our world-view and has destroyed our sense of the meaning and purpose of life. In other words they accuse science of dehumanising and disenchanting our view of the world. Scientific thinking, they say, is 'mechanical' not 'human' thinking and it is high time that we cultivated a more human and spiritual attitude to the world around us.

I can't help remarking, by-the-way, that a lot of these criticisms of the cultural shortcomings of the physical sciences might just as well have been directed against the study of economics. Both economics and the physical sciences see the world in quantitative terms and necessarily exclude from their calculations qualities, such as happiness and kindliness, which cannot be quantified; but it has always seemed to me that it is the economic view of the world which has impoverished our culture more than the scientific. As I see it, the habit of viewing the world in terms of monetary values and of calculations as to whether or not something is 'worth' doing in terms of profit and loss, is more inimical to culture than the mechanical philosophy of the physical sciences; no wonder economics is often called the 'dismal science'.

Throughout history much the same objections to the cultural influence of science have been voiced whenever science has been gaining ground. They are part of an endless struggle to achieve a balance between romanticism and rationalism, between reason and intuition and between the poetic and the prosaic views of the world. Newton himself, whose work contributed more than anything else to the success of the Mechanical Philosophy, feared that too wide an application of that philosophy would disenchant the world by reducing the need for God. Indeed, soon after the great advances made by Newton, there was a strong reaction against the rule of reason, and most of the present-day criticisms of the cultural influence of science were made years ago in the works of Rousseau, Goethe, Blake, Wordsworth and many others. In our own day these same criticisms have been provoked once more by the enormous advances which science has made in the present century, culminating in the dramatic development of the atomic bomb. Their effect has been to make the public less friendly to science, to weaken the ideal of science as a vocation and, generally speaking, to put science on the defensive.

These criticisms of the cultural influence of science have met with remarkably little opposition from any quarter, not only because there is some truth in them, but also because there is little general appreciation of the cultural significance of science beyond its impact on technology. Although the dictionary tells us that science is another word for knowledge and that technology is concerned with the practical application of knowledge, this distinction is rarely made in public discussions, nor is it particularly useful because it is so hard to make in practice. An old joke says that if some venture is successful, such as a landing on the Moon, then the scientists will call it a 'triumph of science', and if it fails they will call it a 'failure of technology'. But this is not what actually happens. Nowadays almost every innovation, from a landing on the Moon to the invention of a better mousetrap, whether it is successful or not, is attributed to 'science'; the word technology is hardly ever used. Thus in its popular image science is inextricably confused with technology and is therefore seen as being primarily an *instrument* for getting new things, new machines, new medicines, but not new understanding. Nowadays it is almost true to say that in most people's minds science is little more than a box of clever tricks which can produce the things which we want; in that sense it is a modern Cargo Cult.

Judging from the fact that very few attempts have been made to defend science against the critics of its cultural influence, this 'instrumental' attitude to science must be shared by many scientists. No doubt many of them feel so secure in the knowledge that their prestige rests firmly on their ability to 'produce the goods', that it is not worth their while to try to persuade people that there is more to science than making a better mousetrap. On the other hand, Gerald Holton [1] may well be right when he writes:

> 'The progress of science to-day is threatened not only by . . . confusion and disenchantment of the wider public. No, what seems to me the most sensitive, the most fragile part of the total ecology of science is the understanding on the part of the scientists themselves of the nature of the scientific enterprise.'

When Adam and Eve ate from the tree of knowledge they were expelled from the Garden of Eden, but they kept their knowledge. As they were leaving the Garden, if we are to believe a cartoon in the New Yorker, Adam said to Eve: 'My dear, we live in an age of transition!'. So it is with us; we live in the transition between an age when knowledge and power were separate to one in which, in Francis Bacon's words, they have 'met as one'. Most of the evils which are blamed nowadays on science are not really part of its nature, but are the consequence of its meeting with power. As Lord Acton observed many years ago: 'Power tends to corrupt.' Like Adam and Eve, we cannot relinquish our new knowledge and return to a state of innocence; instead we must learn to put it to better use. If the first major lesson of the Scientific Revolution was that knowledge is power, the second is that our ability to produce new knowledge greatly exceeds our ability to use that power wisely.

From our discussion two things are clear. Firstly, there is an urgent need to improve the way in which society controls the practical applications of science so that we get more of what we want and less of what we don't want. Secondly, we must try to tell people, including the scientists themselves, not more of science, but more *about* science, so that they no

'Knowledge is Power' . . . 'our ability to produce knowledge greatly exceeds our ability to use that power wisely.'

longer see it in the narrow context of material progress but in the wider context of our whole culture. We shall now look at these two tasks in turn.

### 3.2 **Making better practical use of Science**

Inevitably, some people, because of the many misuses of science, argue that science and technology have their own evil momentum and that, before it is too late, we should give up the whole idea of using them to make progress and return to a simpler more 'natural' life. But as most of us know, that is a policy which is unlikely to work, furthermore it is even more unlikely that any society in the modern world could be persuaded to adopt it! Many of our difficulties, such as pollution and over-population, are indeed due to the application of science, often done with the best intentions, but we must face the fact that our only hope of solving them is to apply more advanced science, or to apply what we already know, more wisely. Our industries, for example, pollute the environment, not because we have used too much science, but because in the past we have used too little; the problems of over-population due to the elimination of disease cannot reasonably be solved by witholding the cures, but they can be helped by developing more effective methods of birth control and agriculture.

Alas the more urgent problem of reducing the threat of nuclear war cannot be tackled in quite the same way. It seems that the most effective thing which most scientists can do is to spell out in gruesome detail the probable consequences of such a war (§3.7). Indeed many of our problems could be helped by more of the sort of information which only scientists can provide.

To get more of what we want and less of what we don't want, we must improve our control over the applications of science as they come to us embodied in new technologies. It is increasingly clear that, to get the best out of these new technologies, we must develop better ways of forecasting and assessing their effects before they are introduced. Many of our current problems such as nuclear power and waste disposal, the use of toxic chemicals, the possible dangers of experiments in genetic engineering, the effects of computers and micro-processors on unemployment, the social effects of new and alien technologies on 'developing countries', are too novel, too complex and too socially pervasive to be controlled simply by legislation or by agencies which are narrowly technical and which are mainly concerned with short-term calculations of only those costs and gains that can easily be quantified.

One classic example of such a problem has been the lengthy debate on the use of the insecticide DDT. Its obvious benefits to agriculture and to the control of malaria can easily be demonstrated and quantified, but there are a large number of secondary effects which are far more obscure. There are, for example, questions about its accumulation in human and animal tissue, its effects on the production of microsomal enzymes in vertebrates, its persistence in the soil and its effect on wildlife. Most of these effects are difficult to demonstrate and quantify; literally thousands of papers have been written about them and the debate involves a much wider range of expertise than chemistry.

There are many other questions, such as the use of *in vitro* fertilisation, which although complex technically, ought not to be settled by people who are expert only in those particular technologies. If we are to avoid a technocracy, then we must first learn to recognize those questions which cannot be answered by science alone, and we must develop better methods by which such problems can be explained and debated in public so that a *wider* range of experience and common-sense is brought to bear on them. As a society we must learn to treat the expert more as a colleague and counsellor and less as an oracle. In his turn the expert must learn to make his subject more intelligible to the layman.

A good example of what I mean is the debate which took place in Cambridge (Massachusetts) in the late seventies when the proposal to construct a laboratory at Harvard for genetic experiments with DNA was brought to the attention of the City Council. After public hearings in which vigorous objections to the proposed work were expressed, the Council appointed a Citizen Review Board to recommend what should be done. The Review Board consisted of eight residents, none of whom were molecular biologists. At one of the hearings, so I have been told, one of the experts was holding forth when a woman member of the Board stopped him and said, in effect: 'Professor, unless you can explain your work in terms which I can understand you are not going to get your building!' He did explain and they did get their building.

The methods and extent to which new applications of science can be controlled more wisely in any particular country must depend, of course, on the way in which its politics are organized and on how social objectives are translated into action. In the USA the Congress established an Office of Technology Assessment in 1972 as an independent advisory body with the job of keeping the Congress informed on technological issues. In Europe, among many other experiments, the OECD has recently been trying to develop methods of assessing the impact of new technologies which, so they hope, can be used by any country. In fact, there is a vast and growing literature on ways in which technical and social considerations can, hopefully, be combined. It abounds with the impressive jargon of the social sciences, but as yet there are very few practical results. No wonder the recent and extraordinarily lengthy report on '*Technological Change in Australia*' [2] quotes the following remark about technology assessment:

'There can hardly ever have been a tool of which we have accumulated so little experience and which has been so extensively analysed in terms of scope, methodology and institutional arrangements!'

In view of this remark I shall not add to the already extensive literature on this topic, but simply note that making better use of the practical applications of science is one of our most urgent and worthwhile problems, and may well prove to be more difficult than actually doing the science itself. However there is one important point about the control of science and technology which I shall now raise, the support of basic research. I have chosen to discuss this topic at some length because, although it is vital to the whole future of science and its practical applications, it is commonly overlooked, or dealt with only briefly, in discussions of how science should be controlled.

### 3.3 **Why support Basic Research?**

It is often said that the real difference between 'basic' and 'applied' research is simply a matter of time scale, and that most research, no matter how recondite, is eventually applied; the more 'basic' it is the longer it takes to prove useful. The abstruse mathematical researches of men like Georg Riemann and William Rowan Hamilton in the 19th century had to wait for the best part of a century before they were found to be useful to modern physics. Nearer our time, Lord Rutherford was fond of proclaiming, loudly, that his work on the structure of the atom in the 1920s was completely useless, and yet the first atomic reactor was operated in 1942!

This argument, that basic and applied research differ only in time scale, obscures the essential difference between their aims. The boundaries between them are not always clear, nevertheless the distinction which I propose to use here is useful and is based on that difference. Basic research aims at increasing our knowledge of ourselves and of the world around us and in the jargon of scientific policy-making is called 'curiosity-oriented'. Applied research is directed towards achieving some recognized practical goal and is said to be 'mission-oriented'.

If now we look at the many official reports and surveys of science policy to see why we should support basic research, the first reason which they give is that it is an investment in the future. They invite us to see it, so to speak, as the seed-corn of the future material benefits which we can confidently expect to flow from applying new knowledge. Indeed many economists [3] have argued that governments must support basic research because the free market will always fail to support it adequately. Hence in a welfare economy, so they say, the government has an obligation to fill the gap between the level of basic research which the market will support and the level which is socially desirable.

It is implicitly assumed in this argument that research is an activity which can be justified in economic terms. At first sight this argument looks plausible, if not self-evident, but in fact it is not easy to prove that it is true for any one country, and that is, of course, the question of most interest to the government and tax-payers of that country. It is all too easy to measure the inputs of time and money into research, but unfortunately it is difficult to measure the output. For example, if we try to measure this output by the rate of growth of productivity, then the statistics for the period 1950-1964 show no clear correlation between the growth of product per head of a country and the amount which it spent on research and development. However we should not, I suggest, attach too much weight to an analysis of such a short period. For example, at that time Japan was importing and licensing much of its new technology. The rapid growth of its GNP did not, therefore, correlate well with its comparatively low expenditure on research and development. However in recent years this situation has changed. In the last few years the Japanese have recognized that without their own indigenous basic research their technology will eventually become out of date. For one thing they have found that countries with advanced technology are not always willing to export their latest ideas to their competitors.

Thus the relation between expenditure on basic research and material progress remains difficult to demonstrate in hard economic terms. As Dan Greenberg has remarked: 'The relationship between research and prosperity is as predictable as that between prayer and deliverance.' As a consequence there is no simple economic index by which a country can measure how much it should spend. Nevertheless, if we look at science as a whole and on a sufficiently long time scale, then we can see clearly the part played by basic research. What we see is not a simple linear process in which research is followed by development and production, but a complex interaction in which the driving forces flow in both directions. Broadly speaking, the social needs and the technology which is available at any given time make possible the application of some particular advance in basic science. But it is, of course, necessary to have already made that scientific advance.

The complexity of these relations are illustrated by the development of the transistor at the Bell Telephone Laboratories in 1947. This development owed a lot to the classic papers on the theory of semi-conductors written many years before, in 1931, by an academic physicist, A.H.Wilson, and also to the more recent advances in the purification of

The relation between the growth of productivity (output per man, OPM) and the amount spent on research and development (R & D). The figures for productivity are delayed by five years to allow time for the R & D to take effect.

germanium. In turn, Wilson's work was one of the first applications of the quantum theory, which itself can be traced back to attempts, early in the present century, to understand the spectrum radiated by a black-body. The quantum theory made use of mathematics which was developed in the 19th century, and so on back into the mists of time.

Another, more quantitative, demonstration of the relations between basic research and applied science is to be found in the study published in 1976 by Julius Comroe and Robert Dripps [4]. In an impressive analysis of the origins of ten major advances in the diagnosis, treatment and prevention of cardiovascular and pulmonary diseases, they found that over 60 per cent of the 529 key scientific articles, on which the advances were based, were concerned with basic research. If the argument, that basic research is essential to the advance of our material progress, is to carry more weight with the public and with the authorities which control the money, it could well be strengthened by more studies of this kind.

A second reason for the support of basic research, commonly found in official reports, is that it contributes to national prestige. To quote a study published by the Science Council of Canada [5]:

> 'The prestige of a nation is to some degree predicated on her contributions of basic knowledge, although such prestige is, to a large extent, restricted to the scientific community.'

This is an argument which I distrust. In times of plenty it can be used to justify a spectacular project – like sending a man to the Moon – in which the expense is not justified by the scientific results, and is therefore likely to get basic research a bad name as being a waste of tax-payer's money. In times of economy it disappears entirely from the scene.

A third reason for the support of basic research, also found in most official reports, is that it builds up a body of expert knowledge and a group of highly trained people who, especially in a small country, are needed to act as an interface between imported technology and local industry, and that it contributes to higher education. This is an important but limited argument; it is limited because many people will claim that there are other more economical ways, such as training programmes, of achieving the same ends.

A fourth reason is that basic research is a valuable cultural activity. This argument suffers from the use of the word 'culture' which, as I said before, conjures up visions of chamber music and folk dancing. As usually expressed it seriously underates the significance of science and presents it as an ornament of society and not, as I shall argue in §3.6, as an activity of central importance to our culture. Let me illustrate this point by quoting from the *Report of the Royal Commission on Australian Government Administration* [6]:

> 'Like the arts', it says, 'Science is one of the graces of life, and its presence as an aspect of a particular society is seen as a mark of civilisation commanding respect from other societies. Since this activity can no longer be performed by the wealthy amateur, a civilised community will, it is argued, properly support it.'

This argument is uncomfortably reminiscent of the reasons sometimes given for supporting a useless and decadent aristocracy!

Finally if we ask the people who actually do basic research why they do it, we shall get all the previous arguments plus the fact that they enjoy doing it, and that they want to find out something new and get the credit for it. If they are working in a branch of science which is far removed from practical application, like astronomy, many scientists, instead of trying to show the relevance of their work to the main body of science, fall back on the 'Mt. Everest' argument – one climbs Mt. Everest 'because it is there' – one studies the Universe simply because it is there. Although this argument may be profoundly true, it is unlikely to impress the average tax-payer, who is more likely to take the view that it is much more fun to climb mountains and look through telescopes than to pay the bills. Generally speaking I have found that scientists engaged on basic research are not much good at justifying their work to the layman.

Thus the conventional apologia for basic science present it as a source of future material progress and of trained people, as a prestigous ornament of society and as a source of personal satisfaction. In my experience these arguments usually fail to carry conviction; they are more likely to be regarded as special pleading by scientists to do what they want at someone else's expense. To be more successful, especially with politicians, they need to be better articulated and to show more clearly the connections between basic research and the practical results in which most people are interested. Even so, I doubt whether they will be really successful until there is a wider understanding of the whole scientific endeavour and of the vital role which it plays in our culture (§3.6).

### 3.4 How shall we choose what Basic Research to do ?

There are some obvious constraints on what basic research we choose to do because science is essentially a social activity and is embedded in history. Inevitably our choice of topics will be strongly influenced by the current preoccupations of society, just as the choice of topics in the physics and mathematics of the 17th and 18th centuries were influenced by the navigational problems of exploring the world, and in our own time the exploration of the solar system has been promoted by the military interest in space research. Our choice will also be strongly influenced by the scientific theories and fashions of the day, by what Thomas Kuhn calls the 'paradigms of science'. A second obvious constraint is that the supply of competent people and money is limited. Nowadays research in most branches of science is so expensive that, if it is to be done at all, it needs a grant from a government, a foundation, an industry or a wealthy patron of science and so, inevitably, basic research is controlled through the supply of money.

This brings us to the first major question in choosing what basic research to do; how much should we spend? As we have already seen, there is no clear-cut economic answer; furthermore, unlike the arts, public appreciation of basic science cannot be measured by attendance at concert halls and art galleries. As far as I can see, at present the amount to be spent is based largely on 'keeping up with the Joneses' – in other words, on looking to see what other countries are doing. This makes the funding of basic research peculiarly vulnerable to changes in the cultural climate and we should aim to make it less so. One of the things we should do is to develop a more convincing case (preferably an economic one!), for the political support of basic research than merely keeping up with the Joneses. We shall also have to pay more attention to the public relations of science. It is just as desirable that the public should understand more about the importance of basic research to science as it is that scientists should understand more about the importance of science to society.

Given some money for basic research, how shall we choose who to give it to and for what? We cannot avoid this choice; to divide the money equally among everybody who would like to do some research is not only impractical but is a sure prescription for mediocrity. We must of course, strike a balance between what I shall call the external and the internal criteria. By external criteria I mean the sort of thing that any good administrator would consider straight away – the size and cost of the project, its suitability to the resources of the particular country, the competence of the people involved, its social importance, and so on. The internal criteria are more subtle. They should, I suggest, favour proposals for research which are not only bright and innovative but which are chosen more for their relevance to the general advance of science than for their relevance to our immediate social needs.

The research proposal which is most relevant to the general advance of science is the one which promises to shed the most light on other topics in science, and thereby advance the frontiers of knowledge at more than one point. Any worthwhile system of funding basic research must recognize that, in the long term, science can only advance on a wide frontier and that it is always difficult, often impossible, to tell what part of that frontier will prove to be most relevant to our social needs, even supposing that we can forecast those needs. It goes without saying that the desire to do something useful for society can be a powerful incentive to support basic scientific research, but it is not a good guide as to what to do. It is unlikely, for example, that the best way to advance our understanding of cancer is to pour huge sums of money into medical research, as was done a few years ago in the USA. Our understanding of cancer may, and probably will, depend on advances on other frontiers of science, such as molecular physics, which are apparently remote from medicine; furthermore the advance of molecular physics will itself depend on advances in other branches of science which are even more remote from medicine, such as mathematics and computing.

We can find quantitative evidence of the value of 'irrelevance' in basic research by looking once again at the study of advances in the field of cardiovascular and pulmonary diseases made by Comroe and Dripps. They found that over 40 per cent of the work on which these advances were based was not clinically orientated at the time it was done. It was in fact aimed more at increasing our general understanding than at any specific application.

This argument that basic research should not be tied to our practical needs is not, as some people suspect, based on a concealed, elitist assumption that basic research is in some way superior to applied research. It is not an argument made from an ivory tower that basic research is a privileged activity which can be justified for its own sake; it is an argument as to *how* basic research can best be done for *all our sakes*. It is a recognition of the historical fact that the future needs of society have been well served in the past by research which was *not* constrained by those needs. Much as it may grieve the tidy-minded people who aspire to plan 'science policies', most efforts to plan research emphasize qualities such as efficiency, accountability and social relevance which are not the qualities which have distinguished the most successful basic research in the past. Indeed it is unlikely that many of the most important discoveries which underlie modern science, such as the atomic nature of matter or the theory of relativity, would have been made by people trying to do 'socially relevant' research; they were made by people who were seeking to know and to understand, not to apply, what they found. As Francis Bacon pointed out: 'Nature to be commanded must be obeyed.' In Bacon's words they were 'obeying' not 'commanding' Nature. Furthermore several of the most socially beneficial discoveries, such as X-rays and penicillin, owed nothing whatever to planning; they were made by accident.

Paradoxically, the popular, often self-righteous and apparently innocuous demand that all research should be relevant to our social needs is one of the greatest dangers to the advance of science, and hence to the long-term satisfaction of those needs. To insist on relevance in basic research is rather like insisting on naturalism in art; if you are successful you end up with something not radically new, but comfortably familiar. Our view of science is still largely anthropocentric – just like the medieval Universe. Copernicus may have removed the Earth from the centre of the scene but he didn't remove us. We still see ourselves firmly in the centre and Nature as being there to serve our needs, and we like to see our scientists doing something useful and socially responsible. One of the dangers of taking this view is that we might forget that we have no real reason to assume that the Universe was designed with our welfare in mind, and that perhaps the worst mistake we can make in our attempts to understand it, is to assume that it was. If the future of science, and of our material welfare, depends on our continued ability to increase our understanding of Nature, then we must recognize that modern physics is telling us that the world is so weird, so apparently alien to our classical ideas about Nature, that it cannot be explored within the narrow context of utility but must be studied on its own terms. As J.B.S.Haldane reminded us, the world may not be only 'queerer than we suppose', it may be 'queerer than we can suppose'. It seems that there is a good deal of truth in what Thomas Huxley wrote in a letter to Charles Kingsley in 1860:

'Science seems to me to teach in the highest and strongest manner the great truth which is embodied in the Christian conception of entire surrender to the will of God. Sit down before the fact as a little child, be prepared to give up every pre-conceived notion, follow humbly to whatever abysses Nature leads, or you shall learn nothing.'

Finally to whom can we entrust the disbursement of money for *basic* research? As we have already seen, the choice of what research we should support involves, among other questions, a judgement of what is most likely to advance the frontiers of science. That is inescapably a scientific matter and must be done largely by scientists. If we give the job to a group of non-scientists they will, especially if they are elderly, probably give most of the available funds to medical research which, like missionary work in the past, they will think of as being good beyond question. They will do this with the best of motives in ignorance of the historical fact that many of the major advances in medical science, such as the discovery of X-rays or the many discoveries of molecular biology, have been made by physicists, and in ignorance of the fact that in

'To insist on relevance in research is to end up with something not radically new but comfortably familiar.' Breakfast in the home in 1985 as foreseen in 1885.

basic research it is the *quality* of the work which is usually more important than the topic, whereas in applied research it is more often the other way round.

Admittedly, the total amount which is spent on basic research is a matter to be decided by the government of a country, but the detailed disbursement of that money can only be done effectively with the help of a group of scientists acting on behalf of the scientific community. If this group is to be up-to-date, they must be working scientists, and if they are to ensure that science advances on a wide frontier, they must be drawn from as wide a range of scientific disciplines as possible. In an ideal world the system would be pluralistic (as it was in the USA for a short time after World War II) and in any one country there would be more than one group making grants. I am, of course, describing a system of 'peer review' which is open to all the well-known criticisms that it is too conservative, that it is closed to public debate, that it is potentially a mutual admiration society and that it wastes the time of working scientists. My own experience of serving one such system – the Australian Research Grants Committee – shows that most of these faults can be largely avoided by simple administrative measures such as limiting the terms of service of any one member, publishing the details of all the grants and by the extensive use of independent assessors of research projects.

No system is perfect, but one which puts the detailed control of basic research largely into the hands of scientists is to be preferred to one which puts it in the hands of politicians or government departments. Almost the reverse is true of the control of applied research. Just as the values of society are not a good guide to the conduct of basic scientific research, so the values of science are not a good guide to its social applications. In the control of applied science we are learning, the hard way, to recognize the responsibility of science to society; in the control of basic research we must recognize the responsibility of society to science.

## 3.5 Learning about Science

At the present time our society attaches great value to the practical applications of science and relatively little to its ideas. This has come about, not simply because these applications have been so successful (§1.13), but also because in the last 100 years the ideas of science have become far too difficult for most people to understand. Physics, in particular, has become so abstract, so highly mathematical and so much at odds with common-sense that, like theology, it is now the province of specialists and, again like theology, has almost ceased to contribute to the mainstream of our culture. It is important, I believe, that all of us should have a wider appreciation of the nature of science and that we should not merely be aware, as at present, of its practical uses and abuses.

Our civilization is often referred to as 'scientific', but that is only true of our gadgets and not of our ideas. To work most gadgets all one needs is to be able to read the instruction book, there is no need to know anything about science. In fact our society is remarkably uninformed about science. The majority of our leaders, politicians, judges, civil servants, religious leaders, editors of newspapers, etc., were students of classics, economics or law. For the most part they learn about science at second hand from journalists and writers of science fiction. Most parliaments, as far as I can see, know very little about science; in the June 1983 elections in the United Kingdom only six scientists, compared to 89 lawyers, were elected to the House of Commons. Indeed the only legislative body which I know to contain a few really competent, albeit elderly, scientists is the House of Lords in Westminster.

As for our universities, the fragmentation of knowledge is now so complete that graduates in humanities are not expected to know any science, and science graduates are not expected to be more than barely literate. If the average graduate in 'arts' were to be transported back in time and asked by an inquisitive Ancient Greek to explain why the sky is blue, why water is wet or glass is transparent, I doubt if he or she would have much to tell. Indeed judging from some recent experiments in the USA on how much university students know about the laws of motion (§2.3), Aristotle would be gratified to find that many of them still share his ideas! If they were also asked to outline some of the principal new ideas which science has produced in the present century or to explain in any detail the role which modern science plays in society, again I suspect that they would not have much to tell an Ancient Greek. However in these days of specialized knowledge we cannot expect, nor is it necessarily desirable, that every 'educated' person should know much actual hard science. On the other hand it is desirable that every 'educated' person should know more *about* science.

Although advances in science and technology have given us an unparalleled ability to communicate with one another by radio, television and vast quantities of print, these so-called 'media' are seldom used to tell us anything about science. They occasionally tell us, breathlessly, about some new 'breakthrough' in medical science, usually a cure for cancer, but we are told little else. And yet many people are interested in learning about science, as is shown, for example, by the remarkable success of the weekly magazine *New Scientist* in the UK or of the weekly radio programme 'Science Show' broadcast by the ABC in Australia. Both of these try to avoid boring people with a lot of science which they probably can't understand, instead they tell people what is happening in science and try to explain its relevance to questions in which they are interested – they try to present science in its social context. And yet there is very little about science seen on the most powerful medium of all, television. There are only too few shows like Richard Attenborough's 'Life on Earth' or Jacob Bronowski's 'Ascent of Man'.

As far as journalism is concerned, it would prefer science to go away; of the 1750 daily papers published in the USA only 50 employ full-time science writers. The reason is, of course, that the popular media prefer topics that are more sensational and entertaining, and so they avoid science; they do this not only because they think science will bore their customers stiff, but also because it is difficult to find people who can put it over to the public successfully. However there is some light at the end of the tunnel. In recent years there has been a marked increase in the number of programmes and semi-popular magazines devoted to science, particularly in the USA, which suggests that, maybe, the popular media have underestimated the market.

The teaching of science in our schools and universities does not do as much as it might to improve this state of affairs. As John Ziman has pointed out in his book on teaching and learning about science and society [7]:

'...the place of science in the popular culture of our time and the role of the scientist in contemporary society are largely determined by the way science is taught in the class-room. Although most people learn very little science, and make very little use of what they learn, they are the silent majority whose views eventually carry more weight than the tiny minority of research workers and advanced technologists. They too must learn something about science as part of their education about things in general.'

And yet that is not the way science is usually taught. Ziman goes on to make the point that most science teaching seems to be aimed at training future professional scientists and is dominated by a rather narrow view of what they need to learn. He argues that it fails to convey the critical, creative and undogmatic spirit of good science; instead he believes that it reinforces some undesirable attitudes to science, in particular the attitude that science is the only authority for belief.

Ziman's main criticism of the teaching of science, that it is too narrow and professional, can be levelled at the teaching of almost any other discipline, law, economics, history, etc. It is a regrettable consequence of the growth of knowledge that all great topics in the Arts and Sciences have been fragmented, especially in our institutions of tertiary education. As I see it, the remedy is, not to try to teach people about science as a separate topic, but to try to teach *all* subjects, including science, so as to show their relevance to knowledge as a whole and to connect them with everyday life.

### 3.6 **The Cultural Function of Science**

To convince people of the practical value of science is only too easy; all we have to do is to point to the latest gadgets in the house or take them through a modern hospital. To point to the fact that it is safe to drink the water out of the tap is not likely to succeed; they will probably take that for granted unless they know something about the history and practice of public hygiene. So it is with the cultural influence of science, unless we know some of its history we take it for granted.

As a start it helps to realize how much the way in which we view and think about the world has changed since earlier times. Before the 17th century our world-view, indeed our whole culture, was closely linked to religion. People looked to religion for the answers to the great questions about life and for guidance as to what was right and what was wrong. As we saw in chapter 1, the Scientific Revolution changed all that, and to-day our culture is linked far more closely to science than to religion. In medieval times it was almost true, to quote Bronowski, ' that it didn't matter what you said as long as it was religious'. To-day it is almost true that 'it doesn't matter what you say as long as it is scientific'. In earlier times people tried to understand the world primarily in terms of meaning and purpose; they asked *why* the Sun was put in the sky and could be satisfied by the answer that it was put there to give us light. Nowadays we ask different questions; we want to know the facts about the Sun, how large and hot it is and *how*, not why, it got to be where it is. One of the major cultural functions of science in our society is to answer these questions and to show us the world *as it is* and not as we would imagine or prefer it to be. By so doing it acts as our essential link with reality and if we fail to maintain this link, then there is no longer any 'nature's truth' nor is there 'public truth'; there is only 'your truth' and 'my truth', and we are in danger of losing the distinction between fact and fiction and between science and magic. The great 18th century philosopher, David Hume warned us of this danger when he wrote:

> 'The imagination of man is naturally sublime, delighted with whatever is remote and extraordinary, and running, without control, into the most distant parts of time and space in order to avoid the objects which custom has rendered too familiar to it. A correct Judgement observes a contrary method, and avoiding all distant and high enquiries, confines itself to common life, and to such subjects as fall under practice and experience; leaving the more sublime topics to the embellishment of poets and orators, or to the arts of priests and politicians.'

However science does not suppress the imagination; much of it, like the arts, springs from the imagination. To advance our understanding of the world scientists first imagine how the world might conceivably be and then, in order to reach a 'correct Judgement', they test their conjectures to see whether or not they 'work', or correspond to reality in the objective world. Artists, on the other hand, test the products of their imaginations to see whether or not they 'work' in the subjective world. History shows us many examples of how dangerous our conjectures about the world can become once they lose touch with reality and are no longer anchored – as is science – to a comparison with observed facts. The trouble is that such conjectures tend to become mirror images of ourselves, often with the worst features exaggerated. We have only to look at the ideas of the Aztecs which led to human sacrifices, or at some of the ideas which underlay the persecution of witches in the 17th century, or at the racial theories of the Nazis, to see what can happen when things are left to 'the arts of priests and politicians'. As Francis Bacon said: 'God forbid that we should give out a dream of our imagination for a pattern of the world.'

But is there, we ask, any good reason to hope that a society which takes its 'pattern of the world' from science will be any better than one that doesn't? Might it not end up as a nightmare, like the others? After all there are plenty of people who say that it will. Some favour a violent scenario with an end in nuclear war, some favour a peaceful scenario with an end in Aldous Huxley's *Brave New World*. There is, of course, no historical evidence to help us answer this question because ours is the first society in history to try to live with the enormous power of modern science, and so our answer can be little more than a personal expression of faith.

As I see it, our best hope of living happily and peacefully in this mysterious world is to try to gain a better understanding of ourselves and of the world around us. To know *what is*, is essential to the solution of all social and moral problems and therefore to any worthwhile vision of progress. Modern science tells us more about ourselves and the world than any society has ever known before. If we are to make wiser use of that knowledge, we must learn to treat science as an integral and valuable part of our culture and not simply as an agent of material progress, and we must accept that the most influential scientific ideas, such as the theory of evolution, are not necessarily the most practical. We must recognize, as Bronowski tells us in the quotation which heads this chapter, that science does have a distinctive and valuable spirit and

'History shows how dangerous our conjectures about the world can become once they lose touch with reality.' The victims of a witch-hunt in the 17th century.

that, if we look a little more closely at the 'pattern of the world' which it offers, we shall see that it opens wider perspectives, fosters important values and leads us into new ways of thinking.

## 3.7 The Perspectives of Science

'A dapper little man but with shiny elbows
And short keen sight, he lived by measuring things
And died like a recurring decimal
Run off the page, refusing to be curtailed'

That is how Louis Macneice saw a scientist. Television shows them to us in compulsory white coats standing beside computers with flashing lights and whirling discs, launching missiles and satellites, or peering sternly down microscopes. They have been brought along to tell us about the latest scientific 'breakthrough' in explanations which we expect to be incomprehensible and loaded with jargon. The more incomprehensible the explanation proves to be, the more we are inclined to believe that the person in the white coat is a genuine scientist; we are inclined to distrust anyone who is too obviously skilled in public relations. This is because in the public image the scientist is an 'expert', a 'back room' boy or girl, whose role is to know a great deal about a very little. This image of a 'scientist' as a specialist with a narrow outlook is a favourite of the critics of science. They like to quote from William Blake:

'May God us keep from single vision and Newton's sleep.'

It is, of course, obvious that any 'single vision' of the world, whether it be scientific, religious or artistic, is likely to be distorted. A view based on the Mechanical Philosophy which includes *only* those features which can be quantified, such as mass and velocity, and excludes qualities which cannot be quantified, such as beauty and happiness, is clearly going to be dismal. As Wordsworth puts it:

Sweet is the love which Nature brings;
Our meddling intellect
Mis-shapes the beauteous form of things:
We murder to dissect.

But in practice people don't look at the world in only one way and most of the scientists whom I know have certainly not lost their appreciation of the 'beauteous form of things', any more than musicians lose their appreciation of music because they dissect it into notes and rules of harmony. To be able to see a rainbow both as a beautiful sight and as an example of optical dispersion, or to be able to read the score of a symphony while listening to it, is more likely to enrich than to 'murder' our appreciation of its 'beauteous form'. The old idea that science constrains the imagination by bogging it down in the facts might have been plausible a long time ago when science was more pedestrian, but nowadays it is hopelessly out of date. Indeed if we look at the history of ideas over the last 100 years it is perfectly clear that it is the discoveries of science, more than any humanistic studies, that have enlarged our view of the world. Modern telescopes have given us an awesome vision of the Universe of galaxies, while X-ray diffractometers have revealed the fantastic complexity of the structure of living matter. The real world which science

'The beauteous form of things', as seen in the 20th century. The Great Nebula in Andromeda, M31.

explores, from the very large to the very small, has turned out to be infinitely more wonderful, complex and interesting than any writer, poet or artist could ever have imagined. Indeed the scientific exploration of Nature enriches our imaginations and is now our principal source of really new ideas. It enriches Art, Philosophy and Religion (§4.6) with new perspectives and ideas, and to Art it gives new techniques. If the Mechanical Philosophy has disenchanted the world, then modern science is well on the way to re-enchanting it.

This new and enchanting picture of the world has shown us our place in time and space in an *evolutionary perspective*. We now see ourselves as an integral part of Nature, made out of the same elements as everything else on Earth and in the sky, and sharing a common ancestry and structure with other living things. In spite of what the so-called 'creation scientists' would like us to believe, astronomers, geologists and biologists have shown beyond reasonable doubt that the Earth and the elements of which it is composed, together with all the life on it, must have evolved over a vast tract of time in which the whole of human history is no more than the tick of a clock. The solar system was formed about 5 billion years ago in a Universe which itself was 'created' some 15 billion years earlier. Primitive forms of life appeared on Earth some 2 billion years ago and human-like creatures have walked the Earth for only a few million years.

This idea of the unity of creation is not new to the world, it is a common feature of Eastern thought, but it has been almost absent from the West for quite a long time and we should welcome it back. It was Descartes in the mid-17th century who dealt it a nearly fatal blow. He insisted that there is an absolute difference in kind between things of the mind *res cogitans* and things in the physical world *res extensa*, and gave us a picture of a human being as a machine inhabited by an immortal and rational soul which dwells in the body and interacts with it through the pineal gland. This dual perspective encouraged a view of the world in which the organic – we should now call it ecological – relation of human beings to their environment was lost to sight and Nature was seen as something to be exploited rather than nurtured.

In the new perspective of modern science our relation to our environment is shown more clearly, and factually, than at any time before; we see that we live on a small planet with limited resources and that, if we want it to support us in comfort, we must take better care of it. Nowadays most scientific studies of the environment must necessarily be based on an international, global, view. To study the pollution caused by acid rain, the effects on the atmosphere of carbon dioxide produced by burning fossil fuels, the destruction of the ozone in the upper atmosphere by the exhausts of high-flying aircraft, the radioactive fall-out produced by atomic explosions and other problems of ecology and conservation, it is essential to consider the planet as a whole, and to learn from science how to analyse global problems quantitatively.

Nothing illustrates more clearly the fact that our most important problems are no longer national, but are now global, than the dangers of nuclear war. Our politicians and military leaders commonly discuss the balance of power between nuclear arsenals, and the possibility of devising a defence against missiles with nuclear warheads, as though they were concerned with an exchange of shell-fire in the 1914-18 war, when it could safely be assumed that the damage caused by a shell would be confined to the target area; they do not tell us what damage might possibly be caused by nuclear missiles outside the target area, to the rest of the world, and to the country which despatched the missiles. The few attempts to answer these questions which have so far been published, suggest that the damage would not be confined to the target area and might endanger life everywhere on Earth.

For example, one recent analysis [8] estimates that in an exchange of nuclear missiles involving less than 40 per cent of the present arsenals of the USA and the USSR, the amount of smoke emitted by the fires into the atmosphere would exceed 100 million tonnes; this smoke would absorb 95 per cent of the sunlight falling on the Earth's surface and the temperature of vast areas would fall below freezing and stay there for months on end. Such a radical effect on the climate might produce all manner of disasters including the failure of crops; there would also be the evil effects of radioactive fall-out which would not be confined to the target area. These calculations are obviously questionable because they involve a number of factors which are uncertain, such as the physical properties of the smoke which would be generated by fires and the circulation of that smoke in the atmosphere; nevertheless that is not the main point. What they show is that the possible effects of a nuclear war must be analysed from a *global* perspective, and that it is the urgent responsibility of the scientific community to make sure that this is done as competently and objectively as possible, and that the results are made known to the public.

Modern science teaches us not only how to see things on a global scale but also how to act on a global scale, and that is a particularly valuable lesson at the present time. Science, as we shall note in §3.8, aspires to be independent of race, colour and creed and in actual fact is the most successful example of international cooperation in the world; it is more successful than religion or sport, being relatively unencumbered by ideology or competition. This international character of science is illustrated by the International Council of Scientific Unions (ICSU) which sponsors research on many such problems and has agencies which study the oceans, the atmosphere, space, the polar regions and many other questions, all from a 'global' perspective. In my own subject the International Astronomical Union is dedicated to international cooperation in astronomy and is supported by all 50 of the countries which are most active in astronomy. It sponsors cooperative research, scientific meetings and the teaching of astronomy all over the world.

No one presents a better example of international cooperation than scientists, particularly astronomers. Astronauts have shown us beautiful pictures in which the Earth is seen from the Moon as a small blue planet in space; we must learn from science to see the Earth in that perspective without going so far away.

Fragile and important as the Earth may look to the astronauts, the astronomers show it to us in a very different perspective. As we learnt in §2.18, there have been dramatic changes in the way we see our world since Copernicus told us that we are not at the centre of the Universe, as we had previously thought, and that everything does not revolve around us and our affairs. The Earth has lost the centre of the stage and is now pictured as a minor planet of an average star in a Galaxy of billions of other stars, and our Galaxy itself is seen to be one of billions of other galaxies which occupy an unimaginably vast space stretching away as far as our largest telescopes can see. Furthermore astronomers and cosmologists advise us that among all those countless stars and galaxies it is probable, almost certain, that there are millions of other planets similar to the Earth on which there may be life. That is, of course, not a new idea; we can find it, for example, clearly expressed in the writings of Cardinal Nicholas of Cusa in the 15th century. What modern astronomy and astrophysics have done is to make it more plausible.

These new ideas of Time, Space and Unity have made a far greater change in our world-view than did the exploration of the world in the 16th and 17th centuries, and in due course will probably have a greater cultural impact. Our current ideas about the meaning and purpose of the world, our philosophies and religions, were mostly formed under the impression that life on Earth is unique and central to the scheme of things, and that everything that can be seen in the sky is relevant in some way or other to the purpose and destiny of mankind. In the new picture the size of the stage on which the drama of human existence is being played and the duration of the performance are both so large as to be beyond comprehension, and the play itself is seen to occupy only a small part of the stage and to have taken, so far, only a few moments of the whole programme. Inevitably these new ideas are reflected in what we think the play is about and how it should be acted. For example, the sheer physical size of this new vision of the Universe, with its vast spaces and billions of galaxies, makes it increasingly difficult to accept that the whole show has been put on solely for our benefit. This is

*"If we don't know how big the whole universe is, then I don't see how we could be sure how big anything in it is either, like the whole thing might not be any bigger than maybe an orange would be if it weren't in the universe, I mean, so I don't think we ought to get too uptight about any of it because it might be really sort of small and unimportant after all, and until we find out that everything isn't just some kind of specks and things, why maybe who needs it?"*

likely to influence, among other things, our ideas about the significance of human life in the Universe, the existence of a personal God, the likelihood of divine intervention in human affairs and the extent to which our future lies in our own hands.

As another example, the enormous increase in the cosmic timescale has given us an almost unlimited past and an almost unlimited future. This 'evolutionary perspective' of the past is influencing our whole attitude to Nature, to the environment and to the importance of change and adaptability in human affairs; while the perspective of an unlimited future makes room for unlimited progress, for a vision of 'men like gods'. Lastly the idea of the Unity of Nature influences not only our attitude to Nature herself but, inevitably, our attitude to our fellow-men. As John Fowles [9] writes in his book on trees:

'Only fools think our attitude to our fellow-men is a thing distinct from our attitude to 'lesser' life on this planet.'

## 3.8 **Science and Values**

Science, we are often told, can tell us only what we *can* do but is no help in telling us what we *ought* to do. Historically this idea was nourished by the philosophy of Descartes which made a cast-iron distinction between things of the mind and things of the body, and also by the more practical need to keep the peace between science and religion. In the early days of modern science in Europe, at the time of Galileo, it was unwise to tread on the toes of the Church which alone could say what *ought* to be done. Robert Hooke, the curator of the early Royal Society, made this very clear in a draft preamble to the statutes of the Royal Society which he wrote in 1663:

'The business of the Royal Society is: To improve the knowledge of naturall things, and of all useful Arts, Manufactures, Mechanick practices, Engynes and Inventions by Experiment – not meddling with Divinity, Metaphysics, Morals, Politics, Grammar, Rhetoric or Logicks.'

The same convenient division of responsibilities was used in the 19th century to pour oil on the troubled waters between science and religion after the storm raised by the theory of evolution. Science, it was agreed, deals with a different, distinguishable kind of knowledge from religion, being concerned only with impersonal facts, it has nothing to say about values.

In this context the word value is one of those hard-working words whose meaning is a bit vague. The *Shorter Oxford Dictionary* tells us, rather high-mindedly, that the ethical meaning of the word value is 'that which is worthy of esteem for its own sake'. In books on moral philosophy the meaning is usually taken to be 'a belief that a specific mode of conduct or objective is personally or socially desirable'. If now, we try to find out from the moral philosophers where our values come from, we get a long and confusing story. Some, like St. Augustine, tell us that they should be based on the will of God and on nothing else; some, like Immanuel Kant, tell us that values are, or ought to be, based on reason aided by what he called 'categorical imperatives'; some like David Hume, tell us that value judgements are based on human nature, on sympathy or selfishness; some, like Jeremy Bentham, tell us that they are based on a calculation of their usefulness in promoting the greatest happiness of the greatest number; more recently, the new and controversial subject of socio-biology claims that it can tell us a good deal about how our values are rooted in the long evolutionary struggle for

Technology in the service of morality.

survival. In the absence of agreement we may take it that we owe our values to a mixture of emotion, self-interest, reason, religious belief and to our experience of living in society both now and in the distant past.

One thing is certain, the idea that we can separate questions of *ought* clearly from a knowledge of what *is* and of what *can* be done has outlived its usefulness, and the belief that science is concerned only with facts and has nothing to say about values is misleading and harmful. It is sometimes used as a convenient excuse, especially by scientifically illiterate politicians and moralistic preachers, to infer that all we need to solve our problems is good-will and the 'right' values, whereas what may also be needed is good science. In fact, one has only to look at the whole host of difficult questions and ethical problems raised, for instance, by genetic engineering, *in vitro* fertilization, birth control, nuclear power, the use of pesticides and other such modern developments to realize that in practical terms it is often difficult, if not impossible, to separate *ought* from *can* – a conclusion which would have pleased Pelagius more than St. Augustine! Experience has shown that wise decisions about what *ought* to be done about many of these problems can only be made by large committees of concerned and expert people in the context of scientific advice as to what *can* be done. Again, a knowledge of what *is* helps to dispel superstition and prejudice. To understand that schizophrenia is more likely to be due to a biochemical disorder of the brain than to possession by evil spirits leads to a more humane and effective treatment of mental illness.

Science has much to contribute to enlightened action; at the very least it can show us how to put our values – wherever they may come from – into practice. As William Blake wrote long ago:

'He who would do good to another must do it in Minute Particulars. General Good is the plea of the scoundrel, hypocrite and flatterer, for Art and Science cannot exist but in Minutely Organized Particulars.'

But there is another, less obvious, reason why the belief that science is not concerned with values is misleading. It treats science as though it were just a collection of data and not a cooperative search for truth which generates its own values. In fact, most people who actually do scientific research find that it is best done by a community which respects certain values. The four principal values, so Robert

Merton [10] tells us in his classic study of the sociology of science, are *Universalism*, *Communalism*, *Disinterestedness* and *Organized Scepticism*. Universalism requires that science should be independent of race, colour or creed and that it should be essentially international. Communalism requires that scientific knowledge should be public knowledge; that the results of research should published; that there should be freedom of exchange of scientific information between scientists everywhere, and that scientists should be responsible to the scientific community for the trustworthiness of their published work. Disinterestedness requires that the results of bona fide scientific research should not be coloured or manipulated to serve considerations such as personal profit, ideology or expediency, in other words they should be honest and objective; it does not, by-the-way, mean that research should not be competitive. Organized Scepticism requires that statements should not be accepted on the word of authority, but that scientists should be free to question them and that the truth of any statement should finally rest on a comparison with observed fact.

Putting all this into less academic language, the actual doing of scientific research encourages people to be intellectually honest, internationally minded, critical of others and yet capable of accepting criticism themselves, and to be ready to publish and to discuss their results.

No doubt this analysis is idealized and out of date. Since it was made the character of the scientific community has changed; it has become larger, more 'industrial', more 'collective' and involved in politics. Nowadays the majority of scientists work in teams and are engaged in military or industrial work; in the interests of secrecy many of them cannot obey all the scientific 'commandments' of an earlier, more academic community. For one thing they cannot put the demands of Universalism and Communalism before those of commercial and military secrecy, and as J.R.Ravetz [11] reminds us in his discussion of the social problems of industrialized science, the competitive pressures of acquiring and distributing funds for research leads to a number of other compromises with the older ideals of science. He suggests, for example, that many of the scientific papers which are published nowadays are intended more to impress the agencies which make the grants than to enlighten the scientific community.

Nevertheless the values of science defined by Merton are, I believe, still personally acknowledged and respected by most scientists. One has only to study the controversy which surrounded Robert J. Oppenheimer and the development of the atomic bomb to realize that one of the problems of maintaining lively and innovative military and industrial research is to reconcile, as far as possible, the personal values of scientists with the quite different values of the organization in which they work.

And so, while it is true that science tries to analyse the world in quantitative, value-free terms, the situation is not that 'science has nothing to do with values', because the actual process of doing cooperative scientific research generates its own set of values. These values are respected by scientists, not because they are personally more virtuous than non-scientists – they seek personal power and glory like any one else – but because they must obey them if they want their work to be accepted and valued by the scientific community.

Needless to say science is not our only source of values, nevertheless it does foster certain values and attitudes which we can ill afford to lose. Perhaps the most important of these is what Merton called Organized Scepticism or, in other words, the pursuit of the 'truth of fact'; a pursuit which, I venture to suggest, tends to make people who engage in it more trustworthy. As I see it, one of the greatest dangers to any society is that it should become too credulous. There are many false prophets and would-be dictators in the world ready to exploit credulity for their own purposes, and the antidote to credulity is a passion for the 'truth of fact'. That may seem to some to be a rather dull, pedestrian, value which needs to be curbed rather than encouraged. Not so! it is one of the most precious and yet vulnerable passions of the human spirit and science is its guardian.

## 3.9 **Beyond the Mechanical Philosophy**

We are always being told in disparaging tones that science is 'mechanical' or 'mechanistic'. This conventional view ignores the fact that much of modern science, especially physics, is now far less mechanistic than it was a hundred years ago. The profound changes which have taken place in the 'philosophy' of the physical sciences should be more widely appreciated than they are because, apart from anything else, they offer us the opportunity of developing better links between science and the other branches of our culture, such as religion and art.

As we saw in §2.4, from the 17th century onwards the mechanical philosophers believed that all nature could be explained in the same way as we understand the working of a machine. They envisaged the world as one great logical system made up of discrete pieces, perhaps atoms, each with its own fixed, intrinsic and independent properties and obeying a few simple laws; basically the problem of understanding the world was reduced to one of dynamics.

This Mechanical Philosophy, widened in the 19th century to include the abstract and non-mechanical notion of fields of force (e.g. electromagnetic fields), was so outstandingly successful in explaining nature and in advancing industry that it became the model, not only for the way we think about the physical sciences, but also for the way we think about the problems of society and of the world in general. Scientific knowledge came to be regarded as the only trustworthy form of knowledge; science was seen as an agency which removed mystery from the world and which, in due course, would explain everything.

Although such a thorough-going mechanistic image of science may have been appropriate to the 19th century it is now seriously out of date. To be sure, scientists would still prefer to be able to explain everything in the world, but after the first flush of enthusiasm and over-confidence in the Mechanical Philosophy they have come to recognize that, although the power of that philosophy to explain things is very great, it is limited. Even their idea of what constitutes an 'explanation' has changed.

As we saw in chapter 2, our exploration of the behaviour of matter on an atomic scale has shown us that the classical picture of the physical world is only an approximation based on our everyday experience of things which we can touch and see. To our surprise we have found that extremely small bodies, such as atoms, behave very differently from the comparatively large bodies we meet in everyday life, and we have been forced to revise many of the well-established ideas of the Mechanical Philosophy which we had come to regard as common-sense. We have learned (§2.15), for instance, that one of the common-sense axioms of that philosophy – that identical causes must necessarily be followed by identical effects – does *not* apply to atomic events and that when we try to predict the behaviour of an individual atom we must exchange certainty for probability. However much we may dislike it, we must accept that, as far as we can tell, the behaviour of individual things on an atomic scale is, in its very nature, unpredictable.

Another common-sense axiom of the Mechanical Philosophy was that the world could be explained, like a watch, by taking it to pieces and that these pieces would have their own intrinsic, independent, properties. In other words an electron or photon would have a 'real', objective, existence just like a stone – a thing with its own properties independent of other particles and of the observer. As Gertrude Stein might have said: 'An electron is an electron is an electron.' One of the great surprises of the present century has been to learn from physics that this common-sense view of our relations with the inanimate world is wrong or, to be kinder, is only an approximation to the truth.

As a final blow to the comprehensive ideas and ambitions of the Mechanical Philosophy, physicists have found that there are some phenomena which cannot be explained in 'mechanical' terms by taking them to pieces. Apparently these events must be treated as a whole, even including the apparatus used to observe them. The EPR paradox discussed under 'A failure of reductionism' (§2.16) is a good example. The fascinating point is that these events are truly mysterious, no one really understands them. Some people say that this is simply a failure of imagination and all we need to do is to change the way we look at the physical world; others suspect that it is a glimpse of the boundaries of the understanding of the nature of matter which science can hope to reach. Only time will tell who is right.

In case what I have just said gives the impression that reductionism as a scientific method is obsolete, I should add that for the solution of most of the problems which we meet it works well enough; however, what modern physics has discovered is that in the atomic world it sometimes fails. The implications of this interesting discovery are not yet clear; it may turn out, for instance, that we may have to scrap one of the basic assumptions on which our attempts to understand the world have always been based, the classical atomic hypothesis that the structure of matter can be understood by breaking it up in a 'linear' process into smaller and smaller pieces. In recent years it has even been suggested by advocates of the 'bootstrap' theory [12] that eventually we shall have to be satisfied with a 'circular' description of matter in which all the so-called 'fundamental' particles are pictured, not as being made up of something even smaller and more fundamental, but as being made up of each other.

### 3.10 **Living with Uncertainty**

If we want to be certain about something, most of us, like Thomas Didymus, prefer to see or feel it for ourselves or, failing that, to know that we could do so if we really wanted to. Most of the factual data on which science is built are like that; they are public knowledge. If we really want to be sure that the polar caps of Mars are white or that bees can detect the polarization of sunlight, we can, at least in principle, go and look through a telescope or study bees. However when we come to the interpretation of these data, to the theories of science, then things are different. As the philosopher of science, Karl Popper, has told us, all non-trivial scientific theories cannot claim to be certainly true; all that they can legitimately claim is that they are consistent with existing observations. In other words scientific theories can never be proved to be certainly true, they can only be proved to be certainly untrue; they are *working hypotheses*.

As an example Joseph Leverrier pointed out in the early 19th century that Newton's laws of motion are inconsistent with the observed precession of the perihelion of the orbit of the planet Mercury because they predict that the precession should be 532 seconds of arc per century, whereas the observed figure is 574 seconds of arc. All efforts to explain this discrepancy failed until, about 100 years later, Einstein showed that Newton's laws are an approximation and are only true in weak gravitational fields and at velocities small compared with the velocity of light. Newton's laws were no longer seen as eternal verities written by God in the Book of Nature, but as 'working hypotheses' written by Newton in his *Principia*. It was seen that their validity had rested on the fact that they worked better than any other laws, until Einstein published his General Theory of Relativity which gave the correct value for the precession of the perihelion of Mercury.

There is a lot to be said for the attitude to certainty which science encourages. The recognition that certainty is something of which we can have very little in this world, and that uncertainty has positive virtues, such as flexibility and open-mindedness, is not only essential to the progress of understanding, but is the foundation stone of tolerance. It is also, as we shall note in the next chapter (§4.6), important to the relations between science and religion. As Oliver Cromwell wrote in a letter to the General Assembly of the Church of Scotland in 1650:

> 'I beseech you, in the bowels of Christ, think it possible you may be mistaken.'

That is exactly what a proper training in the discipline of scientific research should do, teach people to think that they might be mistaken!

### 3.11 **Science as Metaphor**

We all understand what we mean by the word 'real'. A stone is real, not imaginary; it is a solid, dull, inert lump which, if thrown through a window will break the glass. And yet modern science tells us that the inside of this stone is neither solid nor dull; it is mostly space, very peculiar space filled with vacuum fluctuations and 'virtual' particles and in this space there are protons, electrons and so on which sometimes behave as waves and sometimes as billiard balls and which may themselves be made up of other mysterious entities called quarks. Science agrees that our stone is inert, and if we want to throw it back we can predict its path precisely by Newton's laws of motion, or, even more precisely, by Einstein's General Theory of Relativity, but in the scientific description of the stone this apparently dull quality of inertia is itself a fascinating and unsolved mystery. Some scientists think that the inertia of the stone depends upon its interaction with all the other bodies in the Universe! Clearly our concept of a 'real' stone is a simple abstraction from the much more complex properties of stones and expresses our experience of seeing and feeling many stones. It is a *metaphor* which, in terms of our everyday experience, describes something far more complicated, interesting and mysterious.

In the present century physicists have come to realize that because our descriptions of the 'real' world are metaphors based on limited abstractions from a more complex reality, it is possible to arrive at quite different, even contradictory, concepts of the 'thing' which is being observed. A familiar, but not unique, example of this remarkable fact is to be found in the theory of light. As we saw in §2.14, the modern theory of light accepts that it behaves either as an indubitable wave or as an indubitable particle, depending on the type of observation we choose to make. Faced with this strange behaviour we have given up trying to make common-sense out of its properties, and if asked what light is really like, we can only answer, 'light is like light', and offer a mathematical theory which will predict its behaviour in any given situation. The point is, of course, that light is *neither* a particle nor a wave, but something infinitely more complicated, something we cannot visualize in terms of everyday experience. This new view of reality shows us that all our descriptions of objects are not of what they are 'like in themselves', as was envisaged in the Mechanical Philosophy, but are descriptions of how they 'behave' in response to the observations we choose to make. To put it another way, what we call a photon or a wave is more like an 'event' than a 'thing', and there is no longer a clear-cut distinction between the properties of an object and how it is observed. And yet these two concepts, the particle and the wave are both valid within their own limited domain; physicists call them complementary. One of our most common intellectual sins is to confuse a concept with the reality which it represents, and to use it outside its proper domain of validity; in religious language that is the sin of idolatry.

The Theory of Relativity (§2.17) tells us much the same thing about the 'reality' of time and space; it tells us that our common-sense ideas about space and time are abstractions from a more fundamental, complex and mysterious domain which we call space-time. Thus we can think of our whole scientific picture of the physical world in the same way, as a *metaphor* which describes what we observe of a complex, perhaps incomprehensible, reality in terms which we can grasp and use. This picture is limited, not only by our understanding, but also by our tools of observation, so that it is always incomplete, always unfolding and always provisional. It can never claim to be the absolute truth, but at any given time it is the best picture we have.

I hasten to add that the idea that we cannot know what things are 'like in themselves' is not new and can be found in the writings of many philosophers. Immanuel Kant, for example, pointed out in the 18th century that:

> 'Things are given to us as objects of our senses situated outside us, but of what they may be 'in themselves' we know nothing; we only know their appearances....'

What is excitingly new about modern science is that this idea is no longer just one among the many metaphysical speculations of philosophers, but is an idea which has been shown to have significant practical consequences. The phenomena which it helps us to interpret, such as the dual nature of light, can be verified in the laboratory by anyone who cares to do so. Surely it is a valuable step forward to be able to demonstrate to anyone, no matter how sceptical, that our description of the world actually does depend upon how we choose to observe it, and that in consequence there can be more than one valid way of describing the same thing.

This brings us to a profound question about the nature of reality: to what extent does our scientific picture of the world reflect the way we think? May I remind you of Sir Arthur Eddington's story of the ichthyologist who explored the life of the ocean with a net which had a two-inch mesh. He came to the conclusion that all fish are longer than two inches. This little parable prompts us to wonder just how much scientific knowledge is shaped and limited by the structure of our minds? Our experience of physics in this century has made us cautious of answering that question. As we have seen, all the phenomena of nature cannot be explained or described in terms of our familiar, common-sense concepts of space, time, causality, identity or even of locality. The behaviour of objects which are very small, like atoms, or very large, like the Universe, or moving with speeds approaching the velocity of light, cannot be interpreted by common-sense; all these things are radically new experiences of the world, and to bring them within the discipline of science we have had to learn to think in new ways.

A good example is to be found in quantum mechanics where, as we saw in §2.15, we can only predict what an individual particle or proton will probably do by a calculus which involves waves of probability. These waves of probability cannot be detected by any physical observation; they exist, so to speak, only in our minds. Shall we always be able to develop new concepts, such as these waves of probability, to relate and predict phenomena as yet undiscovered? How far can we penetrate the secrets of Nature before our descriptions and explanations become like the words in a dictionary, 'circular' or 'self-referencing'? One of the main articles of faith in the creed of science is that the world is so made that it can be understood by the human mind. To find out to what extent that is true is the great adventure that we call science.

Thus modern physics has shown us that the structure of the physical world cannot be explained in terms of certainty and common-sense as the mechanical philosophers had confidently expected, and that the conventional idea that science will eventually remove all mystery from the world is an illusion. It is true, of course, that science does remove minor mysteries and superstitions, but in doing so it shows us where the major mysteries really are. It has shown us that all our ideas about time, space, particles, light, and so on, are symbols for entities which are fundamentally mysterious and seem to mark the boundaries of scientific understanding. As for the great mysteries which stand in the shadows of all human thought, such as the origin and purpose of the world, modern science cannot be accused of sweeping them away; to answer such questions would require a foothold outside the world, and that is something which science does not have. The mystery of creation is intact, pushed back by 20 billion years or so, but it is still where it always was – in the beginning.

### 3.12 **The Pursuit of Wisdom**

In his vision of a Utopian Society, the *New Atlantis*, Francis Bacon forecast that the pursuit of science would bring,

> '...the knowledge of Causes and the secret motions of things, and the enlarging of the bounds of Human Empire, to the effecting of all things possible.'

Three hundred and fifty years have passed since then and much of his vision has come true, and the continuous advance in our material welfare brought about by the applications of knowledge has, until now, sustained our belief in progress. But now, like Christian in the *Pilgrim's Progress*, we have reached the Hill Difficulty. Our belief in the inevitability of progress through science has been shaken, firstly by the recognition that the power which knowledge brings can be used for both good and evil, and secondly by our many obvious failures to use this new power for our own good. As Bacon foresaw, knowledge has 'met with power'; our problem now is to see that they are joined by wisdom.

One obvious difficulty in making better use of science is that the rapid advance of knowledge in recent years has greatly complicated the problems with which we are faced.

Another less obvious difficulty is that enlightened action requires more than knowledge; it also demands enlightened values and beliefs which, in our present society, have not kept pace with knowledge.

In Francis Bacon's day the values and beliefs of society were to be found, ready made, in the Bible. Science was seen as a game of hide-and-seek in which God first hid the secrets of Nature and then Man tried to find them; the rewards for success were the benefits, mainly material, which science brings. It was quite out of the question that the book of Nature would contradict the Bible, because both had been written by God. The trouble, as we saw in chapter 2, came some years later when people really started reading the book of Nature for themselves and, beginning with Copernicus, found that it did not always agree with the Bible. And so, as a legacy of the Scientific Revolution and the Enlightment, science and religion were estranged and for the past few hundred years have gone their own ways.

As a consequence the systematic expression of value and belief, which in medieval times was the monopoly of religion, has been separated from the systematic understanding of Nature, which is the social function of science. This has happened during a period of the most rapid and fundamental advances in scientific knowledge which the world has ever seen and, not surprisingly, our values and beliefs, to which our religions still make a significant contribution, no longer fit with our scientific knowledge to make a coherent picture of the world.

If our belief in progress is to be further sustained we must, I believe, enlarge our idea of progress to embrace the pursuit of wisdom, and one of the first steps is to recognize that science is an integral part of our culture. In the present chapter I have therefore pointed to some of the ways in which science can enrich our perspectives, values and ideas. In the next chapter I shall argue that another step in that pursuit should be to reconcile the values expressed by our religions with our current scientific knowledge, furthermore I shall argue that we must make use of science to enlighten our religious beliefs.

## References

1. '*The mainsprings of scientific discovery*' by Gerald Holton, in *The Nature of Scientific Discovery*, edited by O.Gingerich, Smithsonian Institution Press, 1975.
2. *Report of a Committee of Inquiry into Technological Change in Australia*, Australian Government Publishing Service, Canberra, 1980.
3. 'The simple economics of basic research', by R.R.Nelson, *J. Pol. Econ.*, **67**, 297-306, 1959: 'Economic welfare and the allocation of resources for invention', in *The Rate and Direction of Inventive Activity*, National Bureau of Economic Research, Princeton University Press, 1962.
4. 'Scientific Basis for the support of biomedical science', by Julius H. Comroe and Robert D.Dripps, *Science*, **192**, 105, 1976.
5. *Background Study for the Science Council of Canada, No.21, Basic Research*, Information Canada, Ottawa, 1971.
6. *The Report of the Royal Commission on Australian Government Administration*, Australian Government Publishing Service, Canberra, 1976.
7. *Teaching and Learning about Science and Society*, by John Ziman, Cambridge University Press, 1980.
8. 'The climatic effects of nuclear war', by R.P.Turco, O.B.Toon, T.P. Ackerman, J.B.Pollock and Carl Sagan, *Scientific American*, **251**, 23-33, August 1984.
9. *The Tree*, by John Fowles and Frank Horvat, Aurum Press, UK, 1979.
10. *The Sociology of Science*, by R.K.Merton, Chicago University Press, 1973.
11. *Scientific Knowledge and its Social Problems*, by J.R.Ravetz, Clarendon Press, Oxford, 1971.
12. 'Bootstrap: a Scientific Idea?', by G.Chew, *Science*, **161**, 762-5, 1968.

# 4 The Religious Dimension of Science

'As one goldfish said to another: "Of course there is a God! Who do you think changes the water?" '

## 4.1 The Divorce of Science and Religion

In their efforts to explore, control and understand the world people have always sought to know not only what there is in that world but also why it is there. To make better sense of what they find or imagine to be true most societies have tried, through scholars and priests, to arrive at a working picture of the world – a 'world-view' – into which they can fit both their knowledge and beliefs. They have put together, so to speak, a scenario within which they can see themselves and their society. In primitive societies where priest and scholar were usually one and the same person, it must have been relatively easy to fit knowledge and belief into a coherent picture, but as knowledge has expanded and as the functions of priest and scholar have become more and more specialized and separate, it has become increasingly difficult. Although this problem is world-wide, I have chosen to limit the discussion in this chapter to the relations between science and formal Christian religion in the western world.

The last widely accepted world-view in which Christian belief and scientific knowledge were successfully brought together in one coherent and comprehensive view of the world was made by the theologians of the Christian Church in the 12th and 13th centuries (§2.1). In those days the job was very much easier for the theologians than it would be nowadays because there were very few scientists to rock the boat and very little scientific knowledge, and so in constructing the cosmology of the Medieval Model the theologians were free to answer many questions which we should now regard as being purely scientific. For instance, while adopting the major features of Greek physics and cosmology, they rejected Aristotle's idea that the Universe had always existed, because it disagreed with the story of creation in the Bible. It is interesting to note that much the same question has been debated in recent years. This time the question has been whether the Steady-State or the Evolutionary cosmology is a better fit to the observational data and, as we saw in §2.19, it looks as though, once more, the decision has gone against the idea that the Universe has always existed.

From a strictly formal point of view, that is to say as a public statement of belief, the Medieval Model was clearly successful and was not in conflict with the science of the day. How successful it was at the deeper levels of the mind is not so clear; a doctrine may be logically invulnerable and yet fail to inspire personal faith. The Medieval Model was com-

Aristotle. Stonecarving from Chartres Cathedral.

pounded of Christian doctrine and Greek cosmology (§2.1). Its complicated system of interlocking spheres of influence was inherited from the Greeks and was really more in tune with the heirarchical structure of medieval society than with the New Testament conception of God and of His relation to Man. As C.S.Lewis [1] has pointed out:

'Delighted contemplation of the Model and intense religious feeling of a specifically Christian character are seldom fused except in the work of Dante.'

Nevertheless the Medieval Model did give western civilization a coherent world-view in which knowledge and belief were combined in a way which was widely accepted and which served a useful purpose for many centuries. In §2.2 we

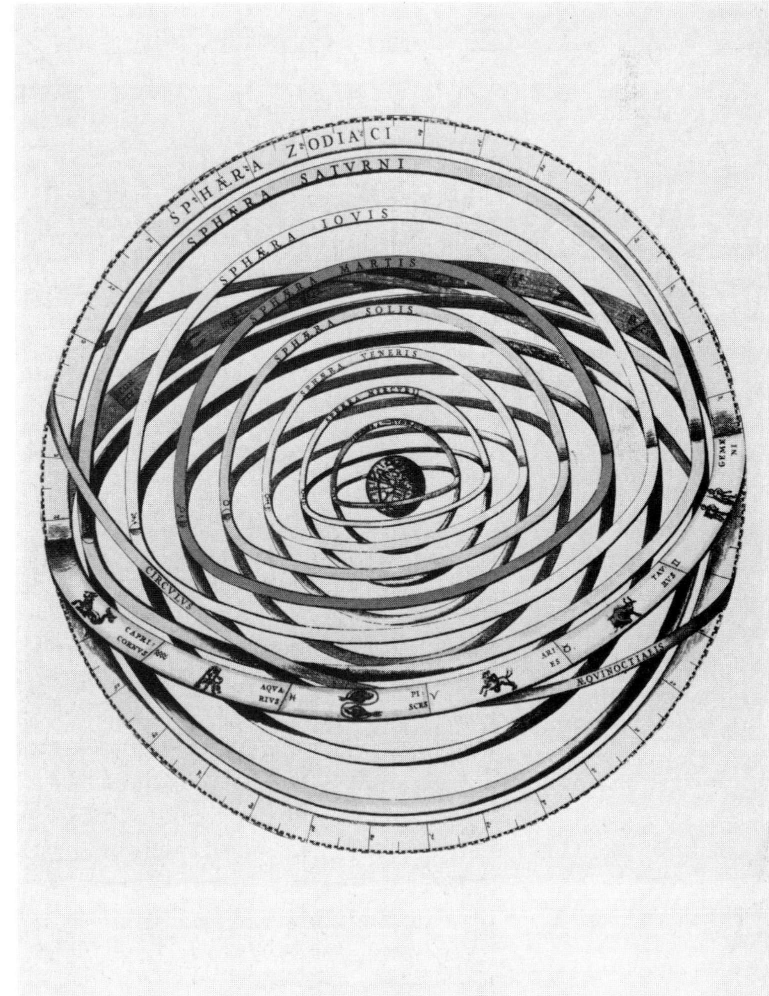


The Ptolemaic universe. The medieval model of the world inherited this complicated system of interlocking spheres from the Greeks.

saw how, from the 16th century onwards, the rise of modern science destroyed this medieval synthesis of knowledge and belief by showing that its cosmology did not agree with what was actually observed in the sky. The Church fought back; it had invested too much of its authority in the Medieval Model to let it go without a fight. At the time of Galileo the Church was powerful enough to silence his advocacy of the heliocentric model of the solar system without open debate, but by the 19th century it had to defend itself against the theory of evolution by public argument. And so there were all those heated – and to us comical – arguments about our descent from monkeys, and about whether or not Adam had a navel and the trees in the Garden of Eden had rings of growth! Broadly speaking science won those battles and the theologians retreated, until by the beginning of the present century the authority of science was vindicated by its practical success both in advancing industry and in explaining nature. It had been established that the physical world could be understood and controlled without invoking God, and the prestige of science grew at the expense of theology.

To be fair we should note that the objections to the religious world-view did not come only from scientists. In the 19th century, during the hey-day of the battles between science and religion, there were volumes and volumes of scholarly 'higher criticism' of the Biblical text; there was Ludwig Feuerbach who told us that the religious view is simply a projection of our own wishes on the sky; and Sigmund Freud who told us that religion is an infantile illusion; and Karl Marx who told us that religion is an ideology of power determined by economic and political interests; and Schopenhauer and Nietszche who told us that our orthodox God is dead, and so on and so on. Almost every conceivable objection to the religious world-view had been articulated by the end of the 19th century, and although they were not all made by scientists it was science which had prepared their way and which took most of the blame.

And what has happened since then? The fighting has died down and for some years now there has been an uneasy truce between science and religion based on the expedient argument, urged towards the end of the 19th century, that they deal with two different kinds of knowledge which are independent of each other. In so far as science has afforded us any alternative perspective to that of religion we may conveniently call it 'scientific materialism'. In this view, as Jacques Monod [2] makes only too clear, reason, observation and experiment are seen as the only trustworthy authorities and scientific knowledge is no longer expected to defer to belief. In the view of scientific materialism God is an unnecessary hypothesis and the idea that there is a God who intervenes in the world is patently absurd; a religion which invokes an interventionist God is seen as a losing cause which no longer speaks to the world in a language which it can understand and which, in due course, will become as obsolete as magic.

This separation between science and religion has been widened by the unparalleled scientific progress made in the present century. Inevitably, many people, under the influence of science, have come to regard the Church as archaic and as something which we can well do without. On the other hand, many people in the Church have come to regard science as the source of our present evils and as an enemy of faith. At best the relations between modern science and the Church are distant. At worst they are antagonistic and present us with two mutually incompatible views of the world.

## 4.2 **The need to reconcile Science and Religion**

Does this separation of science and religion really matter? Alfred North Whitehead [3], a man greatly respected by both scientists and theologians, emphatically thought that it matters a great deal. Fifty years ago he wrote:

> 'When we consider what religion is for mankind, and what science is, it is no exaggeration to say that the future course of history depends upon the decision of this generation as to the relations between them. We have here the two strongest general forces – apart from the mere impulse of the various senses – which influence men and they seem to be set one against the other, the force of our religious intuitions, and the force of our impulse to accurate observation and logical deduction.'

To many people this statement may not sound as significant as it probably did when Whitehead made it, because they associate the word 'religion' only with what goes on in the Church. They assume that, as the influence of the major Churches has declined so has the 'force of our religious intuitions', and in doing so they overlook the many ways in which this force has been redirected outside the Churches in a search for something more satisfactory. Most obvious among the substitutes for conventional western Christianity are the multitude of sects and cults which range from minor variations on the theme of orthodox Christianity to the followers of oriental religions, and from Scientology to a belief in flying saucers which bring us visitors from outer space. Less obvious, but probably more widespread, is the retreat into private and individual religion by people who cannot accept the guidance and interpretation of the Church because, to use Whitehead's words, they find that it sets their 'religious intuitions' against their 'impulse to accurate observation and logical deduction'.

As for the social and practical activities which used to be the special province of the Churches, nowadays there are enough substitutes to suit everybody. Some people, for example, devote themselves to good causes or, indeed, to any activity which involves a belief shared with a group, such as campaigning for social reform or disarmament. For those who feel the need to come together with others to worship, meditate, sing or hear sermons, the Churches are no longer the only places where they can satisfy those needs. If, as the dictionary tells us, to worship is to 'adore with appropriate rites and acts or ceremonies', then there are many ways in which they can do that, ranging from pop concerts to football matches. In the modern world meditation is, I would guess, practised as much outside the Church as in, and as far as listening to sermons is concerned there is no shortage of good advice; we are always being told what we ought to do or think by crusading journalists, talks on the radio or by moralistic plays on television and in the theatre.

For the thinking man or woman who feels the need for a logical structure of belief which does not introduce what Laplace called the 'hypothesis of God', there are the secular faiths of Marxism or Humanism, both of which are expressions of 'scientific materialism'. Marxism is inspired by a vision of a more just and equitable society, and tells us how we can bring it about by a programme based principally on an analysis of the relations between economics and the structure of society. Humanism offers us a vision of human possibilities and a moral purpose inspired by evolution, and teaches us how we can progress towards a better world through the proper use of knowledge and reason. Neither of these secular faiths offers us the assurance that there is a God who cares for the world, nor do they offer us personal immortality; they are both concerned with our welfare in this world and not with our salvation in the next. As compared with Christianity, they tend to locate evil in the social structure and are inclined to see the progress of society as being brought about more by improvement of the organization than by improvement of the individual.

The very existence of these many substitutes for organized religion confirms the continuing strength of the 'force of our religious intuitions', and points to the failure of our major Churches to provide a widely satisfactory social expression of these intuitions. No doubt there are many people who have found what they need in one or other of these substitutes and would not think it worthwhile to bother about what goes on in the Churches; but for the majority these substitutes are not enough; for them private religion is not an adequate substitute for a common faith. Some of us can lead our lives, so to speak, in the presence of an unknown God, but the majority need an organized religion to give a concrete form to their belief and to enrich and support this belief by expressing it in the company of others. The secular faiths, such as Humanism, may satisfy some, particularly those who like to feel that their feet are firmly on the ground; but most people, so the Gallup polls tell us, believe in God, and for them a world without God is rather like Hamlet without the Ghost; a faith based on reason alone leaves them without a sense of meaning and purpose. For most people, reason is necessary but not sufficient. As the scientist, Stephen Weinberg [4], says in his well-known book on cosmology:

'The more the Universe seems comprehensible the more it seems pointless.'

Theodore Roszak [5], a non-scientist and one of the major prophets of the counter-culture, makes the same point more vehemently:

'Our science [like our technics] is maniacal because it bears the cultural burden of finding meaning in our society where meaning cannot possibly be found ....... Nevertheless science continues to thrust its way fanatically into ever denser regions of being, hoping to strike through to some ultimate truth which will vindicate its quest......the secret of life concocted in a test-tube... the origin of the Universe...the mechanism of intelligence...But all it finds are reductionist caricatures, nihilist know-how.'

In the ecology of the human spirit a completely rational argument, like those of the secular faiths, is not enough to inspire and nourish enthusiasm, at least not in the average person. Reason is necessary but not sufficient. There is, it seems, an important part of our minds which moves more comfortably in mystery than in reason; it would rather believe than know and is more at home with ideas that command simple assent than with ideas which require complex understanding or suspended judgement. As Ralph Hodgson [6] puts it:

> Reason has moons, but moons not hers
> Lie mirror'd on her sea
> Confounding her astronomers
> But, O! delighting me.

For most people an assurance that the world has a meaning and purpose is something which religion, not science, must satisfy; for them the scientist can never be an adequate substitute for the priest. Presumably that is what Father Bernard in Iris Murdoch's *Philosopher's Pupil* is trying to tell us when he says: 'Metaphysics and the human sciences are made impossible by the penetration of morality into the moment by moment conduct of ordinary life: the understanding of this is religion.'

The function of organized religion in our society is to give public expression to what Whitehead called 'the force of our religious intuitions', and it cannot do this effectively if it offers us a world-view in which belief is significantly at odds with knowledge. As Aldous Huxley [7] wrote:

> 'Much of the restlessness and uncertainty so characteristic of our time is probably due to the chronic sense of unappeased desires which men naturally religious, but condemned by circumstances to have no religion, are bound to suffer.'

One of the major 'circumstances' which condemn so many of us to have no religion – no formal religion – is that we personally, or the society from which we derive our opinions, cannot reconcile what the priest and scientist tell us about the world. We believe or suspect that science has made religion unnecessary. For that reason it is important that the priest and scientist should get their acts together. A second important reason why we should aim to reconcile science and religion is the vital need to maintain a close link between belief and knowledge. As we noted in §3.6 our ideas about the world, and that includes our religious beliefs, can become dangerous if they lose touch with reality; just as we use belief to illuminate knowledge, so we must use knowledge to illuminate belief.

When the Medieval Model (§2.1) was put together it was possible for one person to know all that there was to be known about science and religion. Nowadays the priest and scientist learn about each other at second-hand and most of their knowledge of each other's profession is gained through the selective and distorting media of journalism, radio or television. One has only to read the proceedings of the conference on Faith, Science and the Future held by the World Council of Churches in Boston in 1979 [8] to see how this mutual ignorance leads to deeply rooted misunderstandings. Our culture would be enriched by a better understanding of our scientists by our priests and vice versa, and by a wider understanding of the nature and function of science and religion in society by all of us.

## 4.3 **The Nature and Function of Science and Religion**

What then, in Whitehead's words, do we consider that 'science and religion are for mankind'? As I see it, the core of religion is the recognition of the *mystery* of existence, of what Rudolf Otto [9] calls the 'wholly other', or, to quote Whitehead [3]:

> 'Religion is the vision of something which stands beyond, behind, and within, the passing flux of immediate things; something which is real, and yet waiting to be realised; something which is a remote possibility, and yet the greatest of present facts; something which gives meaning to all that passes, and yet eludes apprehension; something whose possession is the final good, and yet is beyond all reach; something which is the ultimate ideal, and the hopeless quest.'

At the simplest level it reminds us that the world was not made by man, and that we can't help but wonder how it got to be there and why. Organized religions do their best to interpret this mystery for us by a system of belief which, in the great historical religions (Christianity, Bhuddism and Mohammedanism), is based on what theologians call 'special revelation' in the lives of their founders. Most of them are based on the faith that there is a God who cares for the world and that, through Him, our life has a meaning and purpose, and they assure us that there is life after death. It is the function of our religious leaders and theologians to make these ideas meaningful to contemporary society and, particularly in Western religions, to inspire people to translate them into action.

Science, on the other hand, is an attempt to understand and control this same mysterious world by observation, experiment, reason and working hypothesis. Science is also based on faith, the faith that nature is rational and intelligible and that we can progressively increase our understanding and control of it by free enquiry.

Thus, both science and religion are attempts to interpret the same mysterious world. Organized religion interprets it systematically in terms of the significance of life and relates us to it through awe, reverence, love and ideas of good and evil. Science interprets it for us systematically in terms of knowledge which it aims to make impersonal and objective, and relates us to it through rational understanding and wonder. The vital point which has to be understood, before any progress can be made in reconciling science and religion, is that they are *neither rivals nor alternatives*; in our attempts to make better sense of our experience of a world which is fundamentally mysterious we need them both. Neither of them can claim a monopoly of the 'whole truth' – whatever that may be.

Our problem nowadays, just as Whitehead noted 50 years ago, is that these two interpretations of the world – the same world – seem to be 'set one against the other'. Since his time there have been several attempts to fit them together into one perspective, for example by Teilhard de Chardin [10], and I am sure that there have been many other individuals who have done it to their own satisfaction; but these attempts to reach a world-view acceptable to both science and religion have made little, if any, mark on the mainstream of contemporary thought. In principle the solution is deceptively simple; all we have to do is to recognize which questions can be properly answered by science and which questions by religion. Only by making sure that they both recognize their own limitations can we maintain their mutual respect, and at the same time prevent conflict between these two ways of interpreting this mysterious world.

Pope John Paul II pointed out the importance of recognizing these limitations when he said in his address to the

Commission re-examining the Roman Catholic Church's treatment of Galileo:

'One thus perceives more clearly that Divine revelation, of which the Church is the guarantor and witness, does not of itself involve any particular scientific theory, and the assistance of the Holy Spirit in no way lends itself to guaranteeing explanations that we would wish to profess concerning the physical constitution of reality. It is only through humble and assiduous study that the Church learns to dissociate the essentials of its faith from the scientific systems of a given Age, especially when a culturally influenced reading of the Bible seemed to be linked to an obligatory cosmogony.'

The first thing that happens when these limitations are not recognized is, of course, a conflict of authority. Let us look at the most obvious example, the conflict between science and religion over the authority of Divine Revelation in the Bible.

## 4.4 **Scientific Knowledge and Divine Revelation**

Ever since Anaxagoras was accused of impiety for his teachings about the Sun in 450 BC – and no doubt long before that – most of the friction between science and religion has been caused by the religious claim that truth about the natural world can be guaranteed by the authority of Divine revelation, whereas science claims that it can only be guaranteed by human experience. To see that this conflict is still with us today we have only to look at the clashes between science and the more rigid systems of religious belief, such as the 'fundamentalist' branches of Islam or Christianity, which insist on the central importance and infallibility of the Koran or the Bible.

In recent years there has been a surprising growth of fundamentalist religions particularly in the USA. The Southern Baptists, Latter-day Saints and Seventh-day Adventists, have been gaining ground at the expense of the less literally minded among the Methodists, Episcopalians, Presbyterians, Disciples of Christ and Northern Baptists. The Regius Professor of Hebrew at Oxford, James Barr, tells us [11] that, although fundamentalism has ancient roots in Judaism, it belongs more to the tradition of Protestantism after the Reformation when a stress on the importance of the Bible was part of the anti-Catholic movement; but in its modern form its ideas owe most to the controversy with the rationalists and Deists who attacked the authority of the Bible in the 18th century, and to the 'shock of encounter with

Biblical criticism'. Fundamentalism exists to-day, so Barr says:

'... because there is a powerful tradition in our Protestant religion of Anglo-Saxon countries which can easily lead in that direction. The idea, widespread in our culture, that the Bible and the Bible alone is the sole foundation of religion, is one powerful indicator in that direction. Faced with a variety of currents of modern thought, fundamentalism supposes that it has found a path, based upon the Bible, which offers security and certainty....'

Religion is not, as one might think, to be dismissed as the concern of an unimportant minority. At least one third of the American population now describe themselves as 'born again' Christians and in a recent Gallup poll 44 per cent of people agreed with the statement: 'God created man pretty much in his present form at one time within the last ten thousand years.'

And so, paradoxically, a substantial fraction of the worlds most advanced scientific and technological society accepts the

Reverence for the Bible. Queen Victoria recommends it to an ambassador as the secret of England's greatness (from the British Workman, 1859).

whole Bible as being infallible, and the more literally minded – the 'creationists' – have unfortunately interpreted those parts of it which tell us about the creation of the world as being literally true. They are therefore committed to the belief that they are descended from Adam and Eve, and that God created the world only a few thousand years ago and took only seven days to do it. To believe this they must reject two of the principal ideas on which modern science rests, the theory of evolution and the time scale of modern cosmology.

Contrary to what one might expect, these creationists are not poor and uneducated, many of them come from the technically trained middle-class. However, according to Barr, we should not be surprised that the fundamentalist religions flourish in the centres of high technology because as he points out [11]:

'Fundamentalism is basically an intellectual and rational system, and that is why its power is particularly great in the Anglo-Saxon countries. Its thought patterns go back to the traditional conflicts between faith and rationalism. As against the idea that reason contradicts faith, fundamentalism followed the line of arguments that suggested that reason, properly used, supported faith and indeed demonstrated it.'

Unlike their fellow fundamentalists in the 'developing' countries, who have had comparatively little contact with modern science, the fundamentalists in the 'developed countries', where science is already a powerful influence in their culture, feel that unless science can be reconciled with their faith then that faith itself is threatened. In their efforts to bring about this reconciliation some fundamentalist Churches are trying to remake science by establishing Creation Research Institutes to develop a 'scientific critique' of evolutionary theory and to argue the case for a 'scientific' alternative based on the Bible.

Before we look at what these 'creation scientists' are trying to do, it is interesting to recall that, not so long ago, people were trying to reconcile religion with reason by remaking religion, and it is instructive to see why they failed. Following the breakdown of the medieval world-view (§2.2) there was a vigorous attempt to make religion 'rational' and to base it on 'Natural Theology'. One of the main arguments for 'rational religion', the Argument from Design, was stated forcefully by some of the leading scientists of the 17th and 18th centuries, men like Robert Boyle and John Ray; but its most memorable expression is to be found much later in a book by the Reverend William Paley called by the self-explanatory title *Natural Theology, or Evidences of the Existence and Attributes of the Deity collected from the Appearances of Nature*, 1802. Nowadays most people remember Paley's book for its famous illustration of design. If, as Paley pointed out, we were to come on a watch lying on the ground and were to see how beautifully its parts were made and worked together, we should inevitably conclude that it had been designed for a purpose, and therefore there must have been a designer. By analogy the intricacy and adaptation to their environment of living things, for example of the human eye, implies 'the necessity of an intelligent designing mind' with a purpose in creating the world.

It was this Argument from Design, together with a resentment of the authority of the Church and a wish to combat the use of science to promote atheistic ideas, which encouraged the growth of 'rational' religion in the 17th and 18th centuries. It supported the beliefs of people, like the Deists, who turned away from the Bible, rejected the idea of supernatural revelation as the only guarantee of religious truth, and looked for the revelation of God in nature. To borrow a phrase from a well-known Deist of the 18th century, Matthew Tindal, they saw the Gospels as a 're-publication of the religion of nature' arranged by God for the benefit of those people who couldn't think things out for themselves! The Deists believed that God created the world, set it going, and thereafter did not interfere with the machinery of nature. As a retired engineer His principal functions were seen as twofold, to be the First Cause and to guarantee that there is an absolute difference between good and evil.

These attempts to remake religion on a rational basis soon came to grief. In the latter half of the 18th century the logic of the arguments which supported 'Natural Theology' was attacked and seriously weakened by philosophers such as David Hume, and in the mid-19th century the publication of Darwin's *Origin of Species* demolished the Argument from Design. Rationalists and scientists could no longer rely on Paley's 'evidences', such as the intricacy of the human eye. Science now told them that the eye was not designed; it had evolved by a process in which those variations which were best adapted to the environment survived. It was realized that the appearance of order does not necessarily involve design; it can arise spontaneously by natural processes. Only with the greatest difficulty could such a hit and miss process be seen as being compatible with the creation of man in the image of God as described in the Bible:

'And the Lord God formed man of the dust of the ground, and breathed into his nostrils the breath of life; and man became a living soul.'

Many Christians were prepared to accept that the Bible was not literally true in every detail, but they could not accept that God had designed man in his own image by the haphazard process described by evolution. At what point in this process, it was asked, had man received a soul? Had God created the unfavourable variations as well as the favourable ones which produced man? There were many more trouble-some questions to which there were no satisfactory answers. It seems to me, by-the-way, that the Argument from Design is not completely extinct and, as I have suggested in §2.19, may have reappeared in a different guise as the Anthropic Principle.

This attempt to base religion on reason failed, largely because it was based on arguments which proved to be false, but also because it did not recognize that the primary source of power of living religions is faith, mystery and religious experience and not rational demonstration. When the main bridge between science and religion, the Argument from Design, had been washed away, religion became even more firmly identified with revealed truth and science with 'godless materialism'.

The Creationists, like the Deists, are also seeking rational demonstrations of their faith, but instead of doing this by

remaking religion to fit science, they are trying to do it by remaking science to fit their religion. Consider, for example, the attempts of 'creation scientists' to prove that the Bible story is literally true. The *Book of Genesis* tells us that God created Heaven and Earth in seven days, and the references of the Authorised Version add that He did all this in 4004 BC. This rather mysterious date was computed in 1650 by Archbishop James Usher by adding together the ages of the patriarchs and, according to the 11th edition of the *Encyclopaedia Britannica*, he got it slightly wrong, it should have been 4157 BC!

Modern science disagrees flatly with Usher and puts the age of the Universe at between 10 and 20 billion years, and bases this estimate on a variety of evidence, which includes the rate of expansion of the Universe, the distance between the galaxies, the evolution of stars, the age of the Earth deduced from radioactivity and rocks, and so on. There is so much evidence that must be explained away by anyone seeking to justify the Biblical time scale that it would take a whole book to discuss it adequately.

As an astronomer I was interested to find out the answer to one particular question; how is it that we can see distant galaxies by light which must have taken millions of years to reach us if, as the Bible says, everything was really created only about 6000 years ago? To hear the answer I recently listened to a lecture on 'creation science'. To my surprise the lecturer did not follow the example of the 19th century naturalist Philip Gosse who, faced with the similar difficulty of accounting for the antiquity of fossils, defended the Biblical time scale by claiming that in the beginning God had created everything at the same time, including the fossils.

The lecturer could easily have used the same argument as Gosse; he could have claimed that in the beginning God created the galaxies complete with long beams of light extending from them to Earth. Instead, he was painstakingly 'scientific'. He pointed to the fact that in modern cosmology we assume that the velocity of light has always been the same in the past as it is now, and in so doing, he said, we have all gone wrong. By selecting various measurements of the velocity of light made over a period of several years, the lecturer claimed that the velocity had decreased by 0.5 per cent in the last 300 years. He then fitted a mathematical equation to these selected results and, by extrapolating this equation back 6000 years, came to the remarkable conclusion that the velocity of light in 4000 BC was 500 billion times greater than it is now!! At this enormous speed, light would have travelled some 10 billion light years in the seven days of creation, and therefore Adam, given an adequate telescope, would have seen the same distant galaxies that we can see today.

This argument, presented with all the usual scientific jargon, may sound most impressive to the layman; to a physicist or a cosmologist it carries no conviction. Firstly, if we take *all* the measurements of the velocity of light which have been made over the last 300 years, together with *an estimate of their probable errors*, then they show no significant decrease. Secondly, even if the data taken over 300 years were to show a small decrease, to extrapolate it to yield an increase of 500 billion times in 6000 years is scientifically absurd. I am irresistibly reminded of something St. Augustine said 1500 years ago:

> 'It very often happens that there is some question as to the Earth or the sky, or the other elements of the world...respecting which one who is not a Christian has knowledge derived from most certain reasoning or observation, and it is very disgraceful and mischievous and of all things carefully to be avoided, that a Christian speaking of such matters as being according to the Christian Scriptures, should be heard by an unbeliever talking such nonsense that the unbeliever perceiving him to be as wide of the mark as east from west, can hardly restrain himself from laughing.'

St. Augustine took a fairly dim view of the value of science; even so, no more needs to be said.

The complicated details of these disputes are a matter for specialists and tend to distract attention from the basic reason why the whole approach of 'creation scientists' to science is utterly mistaken. They have failed to grasp the basic point that a religion which demands faith in fixed, revealed truths about nature can never be reconciled with modern science, *not* because science claims to be always right, but because science reserves the right to be wrong and then, if necessary, to change its mind. This is *not* an argument that science is infallible. Science is based on free enquiry; it is a search for truth and in that search it must be free to doubt, to dissent, to enquire and to be wrong. To believe that one has already found the truth is to deny the possibility of progress.

In the 400 years which have elapsed since modern science

emerged from medieval scholarship and escaped from the domination of the Church, scientists have learned that even well-established theories have a limited life. As we saw in the earlier chapters they now know that their theories must be discarded when the progress of science shows them to be either untrue or, at best, to be only approximations to the truth. The advances of 20th century science have shown that our ideas about the nature of the physical world are *not* eternal verities; they are 'working hypotheses' and we should believe them only so long as they work. If scientists were to become, like 'creation scientists', preachers and prophets of one particular set of religious beliefs, then that would be the end of the glorious history of western science; it would have been recaptured by religion and, in due course, would lose both its independence and the support of society.

For that reason any attempt to anchor our beliefs about nature in divinely revealed truth is certain to harm science and, as St. Augustine pointed out, to disgrace religion. Any system of belief, any religion, *must* accept the fact that our knowledge of the natural world cannot be based on the revelation of fixed truth; it must be based on the progressive revelation of truth by experience and scientific research, and for that reason it will always be tentative and can never rest in certainty.

There is absolutely no possibility of fitting what we now know about the evolution of galaxies, stars, chemical elements and of living creatures into the time scale of the Bible. The Biblical story of Creation is a beautiful and instructive myth of which the physical ideas were dissociated from the essentials of the faith by the older Churches long ago. To use the words of Pope John Paul II, the 'culturally influenced reading of the Bible' of those Churches is no longer 'linked to an obligatory cosmogony'. For years they have taught that the Bible story illustrates, not how the world was actually created, but that behind the created world there is a Divine presence and purpose. To teach it nowadays as a literal description of creation is an insult to human intelligence and a grave disservice to our attempts to reach a coherent view of the world. In contrast the theory of evolution, which tells us that all living organisms have descended with change from one or a few original forms of life, helps us to make much better sense of the world. Admittedly it is, like all our knowledge, a 'working hypothesis', and no doubt our detailed understanding of the mechanism of evolution will eventually be modified. In the meantime, the important point is that the theory of evolution works; without it we cannot make sense of biology, and without a time scale of billions of years we cannot make sense of cosmology or astrophysics.

If the creationists persist in their claim that the Bible can tell us how and when the world was actually created, in contradiction to what we are told by modern science, they will only succeed in doing what Whitehead warned us about; they will set the 'force of our religious intuitions' against the 'force of our impulse to accurate observation and logical deduction'. They will further alienate science and religion.

I cannot leave this topic without noting that many fundamentalists believe that the end of the world is not far off, and base their belief on prophecies in the Bible which include the dramatic vision of the Last Judgement given in the *Revelation of St. John the Divine*. To most of us such a belief is faintly comical; it conjures up the vision of an elderly man with a long white beard carrying a placard in the street exhorting us to repent because the end of the world is nigh. My own vision is of an amphitheatre in Sydney, built facing the entrance to the harbour, in which people were persuaded to book seats for the second coming of Christ which was forecast, to the best of my knowledge by Mrs Annie Besant, for some time in 1925. He didn't come, and the site is now occupied by a block of flats.

However what is not so funny is to find that these beliefs are held by people in positions of great authority – even, so we are told, by the President of the USA (Mr Reagan) and some of his principal lieutenants. It is an alarming feature of our present civilization that so many people should embrace beliefs which are grossly at odds with the principal intellectual achievement of our age, modern science. To say the very least, a belief in the inevitable and imminent end of the world is more likely to encourage people to leave things to God than to inspire the patient and constructive planning for the future which the world so urgently needs if it is to avoid, for example, a nuclear war. It rather looks as though a belief in the literal truth of the last chapter of the Bible is a greater danger to society than a belief in the literal truth of the first.

These apocalyptic beliefs illustrate the danger, which we noted in §3.6, of holding beliefs which have insufficient links with reality. Of course science cannot disprove the vision of the end of the world as seen by St. John the Divine; all it can do is to offer its own vision of the future based on what is now known about astrophysics and cosmology. If we look at some of the recent books on cosmology [12] we find that only one feature of that vision seems to be reasonably sure; in about 5 billion years the Sun will evolve into a Red Giant, greatly increase in size, swallow up the inner planets and destroy all life on Earth.

The fate of the rest of the Universe cannot be forecast until we know the total density of matter in space; at present we cannot tell whether it will continue its present expansion indefinitely or will eventually start to contract. In one case, after perhaps 100 billion years, the Universe will become a collection of dead stars (white dwarfs, neutron stars and black holes) drifting away from each other in space; in the other case, everything will contract into an incandescent ball in a reversal of the Big Bang. Either way, the scientific vision of the end of the world is quite as unpleasant as that offered by St. John the Divine. It does, however, have the attractive feature that it offers us an almost unlimited future in which we can aim to be 'men like gods', if only we can learn to survive. It is a vision in which the 'immortality' of the human race depends upon the getting of wisdom.

## 4.5 **Religious Belief and Obsolete Science**

In his speech to the Commission on Galileo Pope John Paul II, having told us that Divine Revelation is no substitute for science, went on to point out that:

'it is only through humble and assiduous study that the Church learns to dissociate the essentials of its faith from the scientific systems of a given age.'

But that is precisely what our Christian Churches have failed to do well enough. To many would-be Churchgoers it seems that 'the essentials of the faith' are still entangled with a web of obsolete ideas, and that much of what the Churches ask them to believe is not only irreconcilable with modern science, but is incompatible with the whole climate of present-day thought. Whatever 'humble and assiduous study' may be going on behind the scenes in cumbersome organizations and Commissions on Doctrine, and whatever theologians write in learned books or the clergy believe in private, it has not yet had enough effect on what actually goes on in our Churches to-day. The beauty and tradition of their rites and ceremonies are an essential link with Christian worship in the past, and may well help people to follow the commandment given by Jesus in Luke 10.27: 'Thou shalt love the Lord thy God with all thy heart, and with all thy soul, and with all thy strength...', but their intellectual content is unlikely to help the educated man and woman to complete the commandment by loving God '... with all thy mind'.

The difficulty is, of course, to ensure that what is actually said in the rites and ceremonies of the Church makes as much sense to the mind of to-day as it did originally, without destroying, not only the link with the past, but also the essential mystery which is always at the heart of worship. There is a need not so much for demystification as for a more contemporarily effective expression of that mystery. The act of worship may have very little to do with the mind, nevertheless the mind must not be offended. The only refuge for the critical mind is to disengage itself from what is being said, and that produces much the same effect as disengaging the emotions from the disasters which we watch during the news on our television sets. The whole thing becomes unreal and, as far as the Church is concerned, there is the danger that a failure to 'believe that' (*fides quae*) will eventually become a failure to 'believe in' (*fides qua*).

To see some of the difficulties which worry ordinary people we have only to look at the many books [13] which have been written, and are still being written, in an attempt to present the teachings of our principal Churches in a way which is more acceptable to the contemporary mind and to persuade us that they can be reconciled with modern science. Some of these books, the simpler ones, start by assuring us that Christian doctrine need not conflict with modern science if only we are prepared, unlike the fundamentalists, to interpret some of the Biblical stories freely. This means, for example, that by treating the story of creation in Genesis as an instructive myth we are made free to believe in the theory of evolution. Again, if we have trouble with the miracles because we find it hard to believe in a God who intervenes in the physical world, then these books invite us to remember that the miracles probably owe their apparently supernatural nature either to a mistranslation of the original text, or to the sort of misunderstanding of natural phenomena which can reasonably be expected of people in those faraway days. For example, the appearance of manna to the Israelites in the desert can be explained in terms of the secretions of an insect (*Trabutina mannipara*); the healing miracles and the casting out of devils can be interpreted in terms of modern pyscho-therapy, and so on.... When all else fails we are invited to treat the miracles as allegories.

To encourage us to interpret the Bible freely is a good start, but most of these books avoid the really sticky questions. For instance, as we were reminded by Pope Pius XII in *Humani Generis* 1951, the Church teaches that we have an immortal soul, but does not make clear when and where that soul entered the story of evolution. It is equally difficult for the man and woman in the laboratory to reconcile the story of the bodily resurrection of Jesus with a scientific disbelief in supernatural intervention in the physical world; why, they may ask, should it not be treated as an allegory, like the other miracles in the Bible? As far as science is concerned the whole idea of bodily resurrection would best be left, once and for all, where it belongs, in those marvellous medieval stained glass windows at Fairford and Chartres, which portray the Last Trump.

Some books go on to tell us how science actually supports the doctrines of the Church, and can even make them easier to believe. For example, modern genetics, so they say, can help us to accept the doctrine of original sin by interpreting it as the inheritance of 'sinful' predispositions in our DNA. Again, if we find it hard to imagine a God who can pay individual attention to millions of us all at once, then one recent book invites us to remember that the theory of relativity tells us that we live in space-time, not in space and time separately, and just as it is possible to be present at one particular point in space at a number of different points in time, so it may be possible to be present at one particular point in time at a number of different points in space! Well-meaning people fill whole books with examples like these.

The general thesis of these books is that the doctrines of the Church make better sense if they are viewed in the right way, and there need be no conflict between them and contemporary science. That may well be so, given enough freedom to interpret the Bible, a fairly considerable knowledge of science and the ability to perform the mental acrobatics demanded by some of their arguments. Nevertheless I am always uneasy about many of these arguments; they seem to me to miss the important point that, in many cases, it is the doctrines themselves which need to be renewed. If the Church had done a better job of 'dissociating the essentials of its faith' from obsolete science there would be no need to write these defensive books. As Jesus pointed out, there are considerable risks in pouring new wine into old bottles. What one really needs are some new bottles.

Although it is always possible to select particular aspects of contemporary science to support established doctrines – for example, the theory of the Big Bang has been used by some theologians to corroborate their ideas about creation – that is not what is really needed. In many cases the doctrines, as they stand, are not worth supporting; they need to be revised in the light of modern science or, in some cases, to be thrown out. No doubt these doctrines were originally hammered out by some of the best minds of their day, but they were made to fit a picture of the world which has long since become obsolete. It was a picture based largely on the physics and cosmology of Aristotle (4th century BC) which was later elaborated into the Medieval Model described in §2.1. In that picture Heaven was completely separate from the Earth, the division being at the orbit of the Moon; things in Heaven were not only made of a different substance from those on Earth but they obeyed different physical laws. Man and the Earth were central to the whole scheme of things; Heaven was above his head and Hell was beneath his feet. Every detail of what Man did and thought was closely watched over by God and His angels.

Too much of what we still hear in the Christian Churches, and too much of what they ask us to believe – certainly too much of what they ask us to say that we believe – is rooted in that old picture. The most obvious example is the concept of Divine intervention in the world, apparently by magic, a concept which was plausible at a time when the majority of natural phenomena, even rainbows, could not be explained by the science of the time. Furthermore there was no reason in those days to assume that God would be limited by the earthly laws of nature, for the very good reason that everything beyond the orbit of the Moon was believed to be literally '*super-natural*'. In such a context it is not surprising that mystery should be interpreted as magic, the essence of magic being that an action is apparently brought about by compelling nature and not by understanding it. That is, of course, what we find in the stories of miracles in the Bible where water is magically transformed into wine and, again, in those doctrines which invoke a belief in magical events such as Virgin birth and bodily resurrection – doctrines to which people are still expected to give public assent when they recite the creeds of our major Christian Churches. Quite clearly the whole idea of supernatural intervention in the physical world, which pervades the traditional presentation of the Christian message, belongs to an obsolete world-view and is hopelessly at odds with the world-view of modern science.

Another example of a concept which was strongly influenced by an obsolete cosmography is the idea of the 'holy' as being something located outside the Earth. In the Medieval Model, God and Heaven were seen as being firmly outside nature and as essentially supernatural, and that is an idea which colours much of what the Christian Church still tells us about the nature of God and the soul. The concept of a physical Heaven and Hell populated by angels, demons and ghosts of the departed, fitted nicely into that early cosmography and, although it may no longer be believed by anyone, it continues to form the background for our image of the soul and for expressions such as the 'Holy Ghost'.

Admittedly our major western Churches have taken some interest in the great changes which modern science has made to our view of the world and particularly in what it has to say about evolution and cosmology; but they have not really

taken these new perspectives of science (§3.7) sufficiently to heart. Whatever our clergy may believe in private, their public expressions of the Christian faith are still presented, to a significant extent, in terms of an obsolete cosmology which has been overtaken by an immeasurably more powerful explanation of the world, the epic of evolution. In comparison with the new and majestic vision of a Universe in which countless galaxies have evolved through aeons of time and in which there may well be billions of other inhabited worlds, much of what the Christian Church asks us to believe as 'essentials of its faith' looks to be centred too closely on Man and the Earth. It is too anthropomorphic and was clearly formulated at a time when people took it for granted that Man and the Earth were unique; as such it is really more suited to the Medieval Model than to modern cosmology. Science is frequently criticized for being too literal and for dehumanizing the world by picturing it too much as a machine; religion can equally well be criticized for being too literal and for humanizing God by picturing Him too much as a person. Both ideas, 'God as a Person' and the 'World as a Machine', belong to an earlier, simpler, stage in the history of thought and have now been overtaken by the development of more abstract ideas.

Religion and science are out of step. In a period of unusually rapid scientific advance, the Christian Churches have been too slow to dissociate the essentials of their faith from the thought, especially the scientific thought, and language of a previous age; this has contributed to an erosion of faith and is one of the many factors which have accelerated the decay of formal religion in western societies. The major growth of Christianity is now taking place in the 'developing' countries. Is this, perhaps, because they are closer to a pre-scientific culture, and so find its doctrines easier to accept?

### 4.6 Religious Belief and Contemporary Science

In his speech on Galileo Pope John Paul II might well have gone on to say that 'dissociating the essentials of their faith from the scientific systems of a given age' is only the negative half of what the Churches should be doing about their relations with science; they should also be taking a more positive interest in contemporary science. Why should they do this?

It has always been recognized that the wonders of nature

*inspire* religious faith, and the revelations of modern science are no exception. Although the day-to-day practice of scientific research as a profession, with its emphasis on objectivity, experiment and precise understanding, may perhaps make people hostile to religious ideas, it is quite a different matter when we come to look at some of the results of that research. For instance our current picture of the Universe (§2.18) certainly inspires wonder, awe and humility, a perspective which is, perhaps, liable to be lost in an irreligious world. Again, modern physics brings us face to face with mystery (§3.11) which, judging from many contemporary books [16], inspires an 'awareness of the transcendent'. Indeed for some people the mysteries of modern physics have proved to be an epiphany. Evidently there is more common 'spiritual' ground on which modern science and religion can meet than most of us realize; but that common ground is not on their periphery, which is what most people see, and can only be reached by penetrating some way towards their core.

Many earnest people, in their efforts to convince others that science can inspire religious belief, point to 'mysteries' on the fringes of science, such as extra-sensory perception and parapsychology and argue that they are proof of an unseen world. In doing so they may well be confusing a problem with a mystery. Whether these phenomena are difficult problems or genuine mysteries must remain sub judice until they have been investigated more thoroughly by scientific research. Not only do these people misunderstand the nature of orthodox science, but they don't seem to realize that because they are dabbling in half-baked science, they are likely to do more harm than good by alienating many scientists. To be sure science and religion do meet in the inexplicable; but, as we have seen, there are plenty of first class mysteries to be found in the well-established and central areas of orthodox science without looking for them on the fringes.

As well as being *inspired* by science, it is only too obvious that a religion which aims to do practical good in the modern world must make *use* of science. If we are to 'love our neighbour as ourselves', then many of our present-day problems require, not simply good intentions, but good science as well (§3.2). We have already discussed how to make better practical use of science (§3.2), and we have seen how its ideas (§3.6), perspectives(§3.7) and values (§3.8) are necessary to enlightened action. In these days, however, when the secular state has taken over so many of the Churches' original social functions, such as education and health, what we need most from our organized religions is a system of enlightened beliefs which satisfies our 'religious intuitions' and which, at the same time, is not at odds with our knowledge. What is not so obvious is that science is as essential to the formation of those enlightened beliefs as it is to enlightened action.

Faith, so St. Augustine told us 1500 years ago, illuminates an understanding of the world gained by reason and is not a mere guessing at undemonstrable truths – '*credo ut intelligam*', not '*credo quia absurdum*', or to put it another way, faith is not 'believing in what you know ain't true'. What the Churches underestimate is the extent to which an understanding of the world gained by reason can *enlighten* their faith.

If we are to appreciate how and why science can enlighten religious belief, we must first recognize that, at heart, science and religion are both attempts to interpret the nature of the same world, and are therefore both ultimately concerned with mysteries which transcend our understanding. Secondly we must realize that the world which they both seek to understand was not made by human beings and that – as religion has always told us and modern science confirms (see §3.4) – it is a world which cannot be fully understood within the limited terms of our immediate practical concerns. This is as true of religion as it is of science, and we can easily lose sight of that fact if we are too preoccupied with the practical applications of either.

Look, for example, at the influence on religion of too great an interest in the practice of politics. In his 1978 Reith lectures Edward Norman [14] criticized the powerful influence which an interest in politics is having on the beliefs of many people in the Church; he called it the 'politicization' of Christianity. Many Christians, so he said, identify the essentials of their faith with their political preferences, and define their religious values according to 'the categories and references provided by the compulsive moralism of contemporary culture'; they see such concepts as 'democratic pluralism, equality, individual Human Rights, the freedom to choose values, and so forth, as basic expressions of Christianity, the modern applications of the moral precepts of Christ'. In Norman's view this politicization of the Christian Church is a symptom of its decay as an authentic religion; in their preoccupation with social and moral action contemporary Christians have tended to lose sight of the most important feature of their faith, which is that it offers the world its own distinctive vision of a meaning and purpose of life transcending any political preferences.

An analogous criticism can be made of the attitude of our society which values science almost wholly for its practical applications – an attitude which in §3.1 I have called a Cargo Cult. Just as an excessive preoccupation with its moral and social applications can lead to decay at the heart of religion, so can an excessive preoccupation with its practical applications lead to decay at the heart of science. As we noted in §3.3 modern physics has shown us that the world is so weird and so alien to our classical ideas about nature that it cannot be explored effectively within the narrow context of the practical applications in which we happen to be interested at any particular moment. A failure to understand the extent to which nature must be studied for its 'own sake' and not for its utility (§3.4), will inevitably arrest the progress of science; if science stands still, it will decay and we shall lose an insight into nature which transcends our concern with its usefulness to us.

Thirdly, if we are to appreciate that science can enlighten belief, we must recognize that it is always important to maintain close links between belief and knowledge, and that it is especially urgent to do so at times like the present when knowledge is changing so rapidly. The major cultural function of science, as we noted in §3.6, is to keep us in touch with reality by continually reminding us of how the world actually *is*, and to do that is *as important to our religious beliefs as it is to any other aspect of our culture*. When religious beliefs lose touch with reality they are likely to turn *inwards* and present a picture of the world which is no more than a mirror of ourselves, and such a picture, as we saw in §3.6, is potentially dangerous. If our system of religious beliefs is to form part of a coherent world-view, as it did in the Medieval Model, it must look *outwards* at what contemporary science is telling us about the world around us.

Fourthly, if as the theologians tell us, Christianity is not 'Man's search for God', but is 'God's search for Man', then it is the obvious duty of the Church to make its message to the man and woman in the street as credible, compelling and intelligible as possible. That necessarily implies that the essentials of faith (whatever they may be) should be conveyed in teachings which are not merely compatible with contemporary science, but which make active use of it. The Bible made use of contemporary science and so did the Medieval Model (§2.1), and there is no reason why the modern Church should not do the same. Both the Bible and the Medieval Model were remarkably effective in their own day and we should not, with hindsight, dismiss their scientific content as ridiculous; in their historical context they were not at odds with what was then known about the world. Once we start to treat earlier beliefs as ridiculous, we are well on the way to regarding our current beliefs as dogma, and anyone who does that is storing up trouble for the future.

What *is* ridiculous is to ask educated people to assent to something which is clearly rooted in obsolete science, and that is what our Churches still do. In every age, old beliefs, as well as their modes of expression, need to be re-examined in

A view of the world before the Space Programme.

the light of contemporary knowledge. It is not enough – although it can sometimes be helpful – to modernize the language of the prayer-book and the Bible by taking out the thee's and thou's, or to load sermons with scientific jargon, while their actual teaching remains incompatible with what we now know about the world. If the Church is to ensure that its message remains compelling and intelligible, then it must keep a sharp eye on modern science. This is not an argument, I hasten to add, that religion should marry the spirit of the age; as Mascall [15] remarks, that can lead to widowhood in the next. It is simply an argument that they should be just good friends.

As an example of modernizing the medium of religion but not the message, it always strikes me as particularly absurd, if not dishonest, that modern evangelists, notably the fundamentalists in the USA, should make use of the whole panoply of communication technology (television, electronic music, etc.) in order to ram home to millions a message which is more often than not hostile to the science on which that technology itself is based!

The Church, like the rest of society (chapter 3) should try to understand what is at the heart of science. Unless it learns to value the ideas of science, and not simply the applications, it will fail to realize that science has much to tell us, not only about enlightened action, but also about enlightened belief.

### 4.7 The Enlightenment of Belief

If, to-day, we were to ask the theologians who put together the medieval synthesis of science and religion (§2.1) to do the same job all over again, what light, if any, would modern science throw on their attempts to reformulate the Christian faith? To start with, what conclusion would they come to about the nature of God? (You have, by-the-way, only to take a glance at the *Summa Theologica* of St. Thomas Aquinas to see that a medieval theologian needed very little evidence to tell the world a great deal about the nature of God!) On taking a good look at contemporary science, and especially at contemporary cosmology, the first thing our theologians would notice would be that in the modern model of the cosmos there is no place for their medieval ideas of a personal God in a separate Heaven and of a personal Devil in a separate Hell, and that in the world-view of scientific materialism there is very little sympathy for medieval ideas about the supernatural. To make a satisfactory synthesis between science and religion, they would be compelled, I suggest, to transpose their old image of a God 'out there' who was in a separate supernatural domain, into a new image of a God who is 'in here' in the domain of nature.

As we have already noted, the doctrines of the Christian Church, like those of most religions, were linked to ideas of the supernatural which were formed in a pre-scientific age, and were given powerful pictorial expression by a cosmology in which there was a clear cut division between Heaven and Earth at the orbit of the Moon (§2.1). This beautiful picture was shattered, as we saw in §2.2, by the progress of science. In particular the work of Newton in the 17th century partly

A view of the world after the Space Programme.

destroyed the cosmological support for the old idea of a separate and radically different supernatural domain by showing that the laws of motion are the same for celestial bodies as they are for things on Earth. The final straw was contributed by the work of Fraunhöfer and Kirchoff in the 19th century which showed that the same substances (sodium, iron, etc.) are to be found in a celestial body (the Sun) as on Earth.

In the last 300 years the support for these early ideas about the supernatural has been further weakened by the astonishing progress of the physical sciences in understanding and controlling nature, until in our own times, the majority of scientists, so I would guess, have no liking for descriptions and explanations which invoke the supernatural. To be more precise they have nothing to say, beyond their dislike, about ideas of the supernatural which can never be tested by observation; but they firmly reject attempts to explain actual events by processes which appear to be contrary to our laboriously and carefully acquired knowledge of the laws of nature or, in other words, by explanations which they believe to be *contra-natural*. As we have seen in chapter 3 there is so much that is wonderful, mysterious and awesome in the 'real' world, that to many people, probably to most scientists, there seems to be no need to invoke the supernatural; the natural is wonderful enough. Must the theologian really try to convince us that a belief in supernatural intervention is one of the 'essentials of Faith', or is it something left over from medieval times when people, even well-educated people, believed in magic? It may be that, for the critical mind, science is an easier road to God than theology!

There seems little doubt that this reluctance of scientists to accept the idea of the supernatural would be one of the main obstacles which our theologians would encounter in their efforts to produce a new synthesis of science and religion. The Bible takes the idea of a separate supernatural world for granted, and so do many Christian teachings, such as those on incarnation, redemption, resurrection and the immortality of the soul; to be sure, many people have interpreted them in 'natural' terms, but that is not how the average Church-goer usually hears them expressed.

Many people will, of course, protest that the Bible and the principal doctrines of the Church cannot be divorced from their original supernatural content without robbing them of their mystery and authority or, in other words, of their holiness and truth. In the light of modern science that argument, I suggest, is questionable. In the present century the explorations of modern science have shown that the natural world is infinitely more vast and wonderful than anyone, however gifted, could ever have imagined a supernatural domain to be. Furthermore, modern physics has shown us, in concrete terms, that as we penetrate further and further into the core of matter our picture of nature becomes increasingly abstract and mysterious; so mysterious in fact that, as we shall note shortly, physicists have been obliged to accept that uncertainty, metaphor and paradox are all inescapable features of our attempts to interpret these mysteries.

In this new perspective there seems to be no compelling reason why we should continue to identify the transcendental with the supernatural; it is also not clear why the doctrines of the Church need lose either holiness or mystery by being removed from an obsolete supernatural domain to take their place in the natural world alongside the mysteries encountered by science. Surely they would gain in credibility, and there would be the positive advantage that the 'idea of the holy' would no longer be exiled outside nature, as it was in the Medieval Model, but would be brought back to Earth where it can do more good. It would also encourage us to base our ideas about God more on what we have discovered about the world and less on what we don't know. The Chinese never identified the holy with the supernatural because, in Chinese thought, there was never anything outside nature. Perhaps we could even reach the happy state when 'Religion and the Occult' is no longer used as single classification as it is in many book-shops, particularly in the USA!

In answer to the protest that the authority of Christian doctrine depends essentially upon supernatural ideas, many of our modern Christian thinkers have argued that it need not. To take but a very few, Tillich [17] has shown us how to picture God as the 'Ground of our Being'; Feuerbach [18] and Robinson [13] have shown us how to preserve the ideas of a personal God and of the divinity of Christ without the support of what Robinson calls 'supranaturalism'. Also to be remembered is that the authority of the Bible, which originally rested largely on 'supernaturalism' because the supernatural was an integral part of the world-view of those days, has declined in modern times because (except for fudamentalists) in the modern world-view that supernatural domain has ceased to exist.

Nevertheless there will always be people who insist that the doctrines of Christianity must be supported by supernatural authority, and that it is the existence of an authority external to nature which is the basic difference between religion and humanism. In that case the important point, as far as the relations between science and religion are concerned, is that any concept of the supernatural which is used should not involve, in any way whatever, the contra-natural. If in their reformulation of Christian doctrine our medieval theologians feel compelled to use the word 'supernatural'and still maintain good relations with science, then they can only use it as a synonym for mystery. The authority of Christian doctrine cannot rest on any specific attempt to interpret the mystery of the nature of God, neither can it rest on any specific interpretation of Nature by science. All such attempts can only be working hypotheses and, as I shall stress later, are inevitably limited by what was known at the time they were made. In due course they are bound to become obsolete and eventually to conflict with contemporary science.

The certainties which people seek in religion can never rest on the shifting sands of metaphysical speculation; ultimately the only secure authority for religious teachings, like the ideas of science, is the *experience*, not the argument, that they are true. In reformulating their central and distinctive doctrine of the divinity of Christ our medieval theologians should take care that it rests on the historical evidence of His life, on the evidence that His teachings work in practice and on the subsequent life of the Christian Church. If they borrow ideas from contemporary science they should use them for illumination and not as foundations.

To persuade the Churches to 'dissociate the essentials of the faith' entirely from obsolete concepts of a separate supernatural domain and to associate them with contemporary concepts of the natural world would, no doubt, be very difficult to do in practice, and in any case is bound to be slow; it is a common feature of debates that one encounters the most firmly held opinions on subjects about which there is the least evidence! Nevertheless that is the direction in which our principal Christian Churches should be moving more rapidly if they wish to remove a powerful source of incredulity which, in an age increasingly pervaded by science, has already alienated many people, particularly those we call well-educated, and will inevitably imperil the future of those Churches themselves.

There are many other insights of modern science, quite apart from its dislike of the supernatural, which would interest our medieval theologians. As one well-worn example, our astronauts have shown us unforgettable pictures of what the Earth looks like from the Moon, and this Gods eye-view has not only converted some of the astronauts into itinerant preachers, but has brought home to many people the fact that we all live on one small planet and that it is our job, as the *Book of Genesis* tells us, to look after it. But the old blueprint: 'Be fruitful and multiply, and replenish the earth, and subdue it: and have dominion over... every living thing that moveth on the earth' is no longer really appropriate. The modern science of ecology tells us in no uncertain terms, that we human beings, like everything else, are dependent on the health and fertility of our environment for our support, and that we are, so to speak, an integral part of nature; to be good stewards of the Earth we must therefore substitute the attitudes of care and cooperation for 'dominion over' nature. The widespread modern concern for protecting the environment is an expression of a recently enhanced awareness – due largely to science – of the interdependence of living things, and of the need to take a more 'organic' and 'global' view of our relations with nature.

An awareness of the unity of nature is far from new. Indeed it is amusing to note that, in his well-known discussion of the history of ideas, Arthur Lovejoy [19] lists it as one of the seductive ideas to which philosophers are particularly susceptible. He points out that to say that 'All is One' engenders a congenial mood in which we feel a welcome sense of freedom and relief from the limitations of individuality, and that 'to recognise that things which we had hitherto kept apart in our minds are somehow the same thing ... is normally an agreeable experience for human beings'. Nowadays we can succumb to the temptation of this congenial belief without a feeling of guilt because the recognition of the interdependence of living things is no longer at the mercy of moods or the fashions of philosophy; it can be demonstrated in detail by ecologists. Our western religions can therefore welcome back this insight with the blessing of science; it is an old friend which, in earlier times, illuminated their faith with what Thomas Goldstein calls [20]:

'The lovely Medieval idea of *ordo mundi*, the faith in a universal order, a religious feeling for the ultimate unity of life.'

As yet another example of how contemporary science can illuminate belief, we saw in chapters 2 and 3 that our modern physicists have some interesting things to say about the limitations of knowledge, and much of what they have to say is of great importance to the relations between scientific and religious thinking which are largely concerned with those limitations. Observations of the behaviour of matter on an atomic scale have made it clear that our scientific picture of the world is not, as classical science believed, a picture of what things are 'like in themselves' but is a picture of our experience of relationships and events; it is a *metaphor* (§3.11). *All* our attempts to describe the world, from the electron to the Trinity, are only useful metaphors and must be discarded when they no longer work. We can only describe the unknown in terms of the known and so, as the frontiers of knowledge enlarge, our metaphors must change.

Physicists have also exposed the extent to which our knowledge of the world can be said to be certain. In the atomic world we have been forced to exchange probability for certainty (§2.14, §3.10), and to recognize that in answer to many of our questions about this mysterious world the best that we can hope for is a satisfactory working hypothesis. Again it must be recognized – and this is difficult for many people to accept – that *all* our non-trivial interpretations of the world, religious and scientific, are more or less uncertain, they are working hypotheses, and like metaphors must be discarded when they no longer work. In the words of Aldous Huxley [21]:

'...one of the great discoveries of modern times is the working hypothesis, which has replaced the idea of the dogma or the doctrine.'

Scientists have learned, like Descartes, that there is not much in this world about which we can be certain; not unreasonably most of them suspect that to be even more true of an unseen world.

Both these ideas about the limitations of knowledge can help us to improve the relations between science and religion. To expose the metaphorical and hypothetical nature of scientific knowledge is to weaken the idea that the only valid way of relating to the world is through objective knowledge, and by so doing to make more room for other ways of 'knowing'. For instance, to recognize that our picture of the world is an account of how things act and not of what they are 'like in themselves', makes more room for the familiar statement of the religious experience that God acts in the world by being believed in [22]. They are also a warning to hard-headed scientists that they may have to follow the advice given by Hamlet when Horatio found the Ghost 'wondrous strange'. Hamlet said:

'And therefore as a stranger give it welcome. There are more things in heaven and earth, Horatio, than are dreamt of in your philosophy.'

On the other hand these insights of physics are also a warning to the theologians that their elaborate speculations about the nature of God, for instance the doctrine of the Trinity, are no more than working hypotheses and are useful only so long as they 'work', or in other words, so long as they represent and illuminate our experience of God.

As we noted in §3.11 these 'new' ideas about the limitations of knowledge are in fact 'old' in the sense that we can find them in classical philosophy; but they are 'new' in the important sense that it is only in the present century that scientists have shown them to be, not just interesting metaphysical speculations, but ideas that actually work in what we like to call the 'real world'. Furthermore these ideas are not as foreign to religious thought as many people would imagine. As John Hapgood, the Archbishop of York [23], writes:

'To admit the relativity of all merely human formulations is the most profound way of acknowledging the ultimate authority of God.'

Another important lesson from science, which modern physics illustrates so beautifully, is that in our attempts to describe mystery we must not be surprised if we are compelled to use *paradox*. The physicist, faced with explaining experiments on the nature of light, reluctantly talks about the two-fold nature of the photon; the Christian theologian, faced with explaining religious experience of the nature of God, talks about the three-fold nature of the Trinity. Both these descriptions should be recognized for what they are, attempts to describe our experience of a mystery. There is, however, one vital point which modern physics makes very clear and which cannot be overemphasized; every metaphor or symbol which we use to describe our experience of the world is based only on those limited aspects which we choose to observe. If we choose to do experiments in which we observe the interference of two beams of light, then we may usefully describe light as a wave; if on the other hand we choose to do experiments in which we observe how it gives up its energy to a detector, then we may usefully describe light as a particle. The explanation of this paradox is that both of these apparently contradictory descriptions, particle and wave, are rigorously valid, but they are only valid in the limited context of the appropriate experiment; outside that context they may well give a wrong, even absurd, answer. Much the same thing is, I suggest, true of the theologians' attempts to describe the nature of God.

We must always remember that *the metaphors and symbols which we use to describe our experience of the world are only valid within the limited context of that experience and cannot be used outside it*; they are useful simplifications of the world and must never be confused with the much more complex reality which they represent. The confusion of a symbol with reality is, as I have said before, idolatry, and that is a sin to which we are all prone. Indeed many of the difficulties which prevent ordinary people from reconciling science with religion are, as we have already seen (§4.4), due to a failure to recognize the context within which the metaphors and symbols of science and religion are valid. For example, the old arguments about transubstantiation and the modern arguments about bodily resurrection illustrate the dangers of using symbols outside their proper context; it should be made clear to the man and woman in the pew that the proper context of most religious symbols is not the world of things which we can touch and see.

One of the most profound lessons to be learned from science is not what it has done, but the way it goes about doing it. In its pursuit of truth science welcomes change and accepts uncertainty as inevitable; it is a progressive adventure into the unknown, and in any genuine adventure there is always a risk of being mistaken. Unless our Churches can learn to see religion in much the same spirit, then, like many other religions before them, they will pass slowly into oblivion and their place will be taken by other, perhaps more adaptable, faiths. Religion as a part of society will surely endure, because for most people the choice will not be between having or not having a religion, it will be between one religion and another.

If our Christian Churches are to give our society what it needs, a system of enlightened beliefs which is not at odds with contemporary science and which will lead us to a more coherent view of the world, then they should actively welcome science as an ally. They should recognize that although a superficial knowledge of science may sometimes encourage irreligion, a deeper knowledge does not. As we have seen, modern science tells us that we are an integral part of nature, dependent on cosmic forces beyond our control, and that time, space, matter and perhaps consciousness are mysteries which transcend our present understanding. That is a view of the world which is consistent with the basic religious intuition that Man is not the measure of all things.

Pythagoras. Stonecarving from Chartres Cathedral.

Indeed the Churches would do well to view the remarkable new understanding of nature which science has brought, not as an attack on religion, but as an assurance that we take part in the nature of God.

We are often told that: 'The fear of the Lord is the beginning of wisdom'; but very seldom do we hear the rest of the quotation, 'and the knowledge of the holy is understanding'.

Our Christian Churches have allied themselves with science at times in the past and that is the direction in which they should now be moving. If they want a precedent and a practical example of how it can be done, they have only to look back to the cathedral-school at Chartres where, in the 12th century, the sciences were taught alongside theology and the liberal arts; in fact the sciences were, at least for a time, given precedence over the other liberal arts. The aim of one of the great teachers at Chartres (William of Conches), so Thomas Goldstein [20] tells us, was:

> 'to replace the Other-worldly universe of the conservative theologies with a new natural cosmos that was to be equally complete and at the same time This-worldly.'

Over the great west door of Chartres Cathedral are the images of four great scientists (Euclid, Pythagoras, Ptolemy and Aristotle) together with Christ and the Saints. They were put there in the 12th century as an expression of the unity of science and religion. Can the Christian Church, eight centuries later, recapture that vision?

## References

1. *The Discarded Image*, by C.S.Lewis, Cambridge University Press, 1964.
2. *Chance and Necessity*, by Jacques Monod, Collins, Glasgow, 1972.
3. *Science and the Modern World*, by A.N.Whitehead, Cambridge University Press, 1926.
4. *The First Three Minutes*, by Steven Weinberg, Andre Deutsch, 1977.
5. *Where the Wasteland Ends*, by Theodore Roszak, Faber and Faber, London, 1973.
6. *Collected Poems*, by Ralph Hodgson, Macmillan, London, 1961.
7. *Proper Studies*, by Aldous Huxley, Chatto and Windus, London, 1927.
8. *Faith and Science in an Unjust World*, World Council of Churches, Geneva, 1980.
9. *The Idea of the Holy*, by Rudolf Otto, Oxford University Press, 1923.
10. *The Phenomenon of Man*, by Teilhard de Chardin, Collins, London, 1959.
11. 'Religious fundamentalism', by James Barr, *Current Affairs Bulletin*, **59**, 4-30, 1982.

12. For example, *Cosmology*, by E.R.Harrison, Cambridge University Press, 1981: *The ultimate fate of the universe*, by J.N.Islam, Cambridge University Press, 1983.
13. For example, *But that I can't believe*, by J.A.T.Robinson, Collins, London, 1963: *Honest to God*, by J.A.T.Robinson, SCM Press, London, 1963: *Science and the Renewal of Belief*, by Russell Stannard, SCM Press, London, 1982: *Priestland's Progress*, by Gerald Priestland, Ariel Books B.B.C., London, 1981: *The Way the World Is*, by John Polkinghorne, SPCK, London, 1983.
14. *Christianity and the World Order*, by Edward Norman, Oxford University Press, 1979.
15. *Christian Theology and Natural Science*, by E.L.Mascall, Longmans, London, 1956.
16. *The Tao of Physics*, by Fritjof Capra, Fontana, 1975: *Physics as Metaphor*, by R.S.Jones, University of Minnesota, 1982.
17. See for example the discussion in *Courage To Be*, by Paul Tillich, Fontana, London, 1952.
18. *The Essence of Christianity*, by L. Feuerbach, 1854, quoted in *Honest to God*, by J.A.T.Robinson in ref.12
19. *The Great Chain of Being*, by A.O.Lovejoy, Harvard University Press, 1936.
20. *Dawn of Modern Science*, by Thomas Goldstein, Houghton Mifflin, Boston, 1980.
21. *The Human Situation* by Aldous Huxley, Chatto and Windus, London, 1978.
22. 'Faith, science and the future; the conference sermon', by John Hapgood in *Faith and Science in an Unjust World*, Vol.1, World Council of Churches, Geneva, 1980.
23. *Church and Nation in a Secular Age*, by John Hapgood, Darton, Longman and Todd, London, 1983.

# Selected Bibliography


**Histories of Science**

*The Scientific Revolution, 1500-1800*, by A.R.Hall (Longmans 1954).

*Augustine to Galileo*, Vol.1 *Science in the Middle Ages*: Vol.2 *Science in the Later Middle Ages and Early Modern Times*, by A.C.Crombie (Heinemann 1957).

*Thomas Sprat's History of the Royal Society*, eds. J.I.Cope and H.W.Jones (Routledge, Kegan Paul 1959).

*Science in History*, Vol.1 *The Emergence of Science*: Vol.2 *The Scientific and Industrial Revolution*: Vol.3 *The Natural Sciences in Our Time*: Vol.4 *The Social Sciences: conclusion*. By J.D.Bernal (Penguin 1965).

*A Short History of Science*, by J.D.Crowther (Methuen 1969).

*Victorian Science*, by George Basalla (Doubleday 1970).

*The Dawn of Modern Science*, by Thomas Goldstein (Houghton Mifflin 1980).

**Histories of Ideas**

*Science and the Modern World*, by A.N.Whitehead (Cambridge University Press 1926).

*The Idea of Progress*, by J.B.Bury (Macmillan 1932).

*The Making of the Modern Mind*, by J.H.Randall (Houghton Mifflin 1940).

*The Origins of Modern Science*, by H.Butterfield (Macmillan 1949).

*The Evolution of Scientific Thought*, by A.d'Abro (Dover 1950).

*The Great Chain of Being*, by A.O.Lovejoy (Harvard University Press 1956).

*The Copernican Revolution: Planetary Astronomy in the Development of Western Thought*, by T.S.Kuhn (Harvard University Press 1957).

*Short History of Scientific Ideas to 1900*, by Charles Singer (Oxford University Press 1959).

*The Mechanisation of the World Picture*, by E.J.Dijksterhuis (Oxford University Press 1961).

*The Structure of Scientific Revolutions*, by T.S.Kuhn (University of Chicago Press 1962).

*The Western Intellectual Tradition*, by J.Bronowski and B.Mazlish Penguin 1963).

*Conjectures and Refutations: The Growth of Scientific Knowledge*, by K.R.Popper (Routledge, Kegan Paul 1963).

*The Discarded Image: an introduction to Medieval and Rennaisance Literature*, by C.S.Lewis (Cambridge University Press 1964).

*The Philosophy of the 16th and 17th Centuries*, ed. R.H.Popkin (Macmillan 1966).

*A History of Medieval Philosophy*, by F.C.Coppleston (Methuen 1972).
*Darwinian Impacts*, by D.R.Olroyd (University of New South Wales Press 1980).

**General Books about Science**

*The Common Sense of Science*, by J.Bronowski (Heinemann 1951).
*The Edge of Objectivity*, by C.C.Gillespie (Princeton University Press 1960).
*Science since Babylon*, by D.J. de Solla Price (Yale University Press 1961).
*Little Science, Big Science*, by D.J. de Solla Price (Columbia University Press 1965).
*What is Science for?*, by Bernard Dixon (Collins 1973).
*Science Observed*, by F.R.Jevons (George Allen and Unwin 1973).
'Mainsprings of scientific discovery', by Gerald Holton in *The Nature of Scientific Discovery*, ed. O.Gingerich (Smithsonian Institution Press 1975).
*The Force of Knowledge: the Scientific Dimension of Society*, by John Ziman (Cambridge University Press 1976).
*Reliable Knowledge: an exploration of the grounds for belief in science*, by John Ziman (Cambridge University Press 1978).
*Science, Ideology and World View*, by C.C.Gillespie (Princeton University Press 1981).

**Science in relation to Society**

*The Social Function of Science*, by J.D.Bernal (Routledge 1939).
*Science and Human Values*, by J.Bronowski (Harper 1956).
*Modern Science and Human Values*, by E.W.Hall (Van Nostrand 1956).
'The emergence of science as a profession in nineteenth-century Europe', by Everett Mendelsohn in *The Management of Scientists*, ed. Karl Hill (Beacon Press 1964).
*The Deluge: British Society and the First World War*, by Arthur Marwick (Bodley Head 1964).
*Science and Society*, by Hilary and Steven Rose (Penguin 1970).
*The Making of a Counter Culture: Reflections on Technocratic Society and its Youthful Opposition*, by Theodore Roszak (Faber 1970).
*Sociology of Science*, by R.K.Merton (Chicago University Press 1973).
*Where the Wasteland Ends: Politics and Transcendence in Post-Industrial Society*, by Theodore Roszak (Faber 1973).
*On Human Nature*, by E.O.Wilson (Harvard University Press 1978).
*The Reenchantment of the World*, by Morris Berman (Cornell University Press 1981).
*Science and Social Change*, by Colin Russell (Macmillan 1983).

**Science and Technology**

*Technology and the Academics: an Essay on Universities and the Scientific Revolution*, by Eric Ashby (Macmillan 1958).
*Short History of Technology*, by T.K.Derry and T.I.Williams (Oxford University Press 1960).

*The Unbound Prometheus: Technological Change and Industrial Development in Western Europe from 1750 to the Present*, by D.S.Landes (Cambridge University Press).

*Science, Technology and Economic Growth in the Eighteenth Century*, by A.E.Musson (Methuen 1972).

*Research and Technology as Economic Activities*, by Kenneth Green and Clive Morphet (Siscon, Butterworths 1977).

**Astronomy and Physics**

*Causality and Chance in Modern Physics*, by David Bohm (Routledge, Kegan Paul 1957).

*The Fabric of the Heavens*, by Stephen Toulmin and June Goodfield (Hutchinson 1961).

*The Architecture of Matter*, by Stephen Toulmin and June Goodfield (University of Chicago Press 1962).

*The Conceptual Development of Quantum Mechanics*, by Max Jammer (McGraw Hill 1966).

*Modern Cosmology*, by D.W.Sciama (Cambridge University Press 1971).

*The Philosophy of Quantum Mechanics: the interpretations of quantum mechanics in historical perspective*, by Max Jammer (Wiley 1974).

*The Tao of Physics*, by Fritjof Capra (Fontana 1976).

*Cosmology: the Science of the Universe*, by E.R.Harrison (Cambridge University Press 1981).

*The Ultimate fate of the Universe*, by J.N.Islam (Cambridge University Press 1983).

**Religious Topics**

*The Idea of the Holy*, by Rudolf Otto (Oxford University Press 1923).

'The substitutes for religion', by Aldous Huxley in *Proper Studies* (Chatto and Windus 1927).

*Christian Apologetics*, by Alan Richardson (SCM 1947).

*The Shaking of the Foundations*, by Paul Tillich (SCM 1947).

*Religion and the Scientists*, ed. Mervyn Stockwood (SCM 1959).

*Honest to God*, by J.A.T.Robinson (SCM 1963).

*Religion and the Decline of Magic: Studies in Popular Beliefs in 16th and 17th century England*, by Keith Thomas (Penguin 1971).

*Religion and the Rise of Modern Science*, by R.Hookyas (Scottish Academic Press 1972).

*Church and Nation in a Secular Age*, by John Hapgood (Darton, Longman and Todd 1983).

*The Way the World Is*, by John Polkinghorne (Cambridge University Press 1983).

**Historical and Biographical**

*The Crime of Galileo*, by Giorgio de Santillana (Heinemann 1958).

*Francis Bacon, from Magic to Science*, by Paolo Rossi translated by Sacha Rabinovitch (Routledge, Kegan Paul 1968).

*The Advancement of Learning*, by Francis Bacon ed. Arthur Johnston (Oxford University Press 1974).

*Descartes: Critical and Interpretive Essays*, ed. Michael Hooker (John Hopkins University Press 1978).
*Never at Rest: a biography of Isaac Newton*, by R.S.Westfold (Cambridge University Press 1980).
*Subtle is the Lord: the Science and Life of Albert Einstein*, by Abraham Pais (Oxford University Press).

# Index